Early Trench Tactics in the French Army

The Second Battle of Artois, May–June 1915

JONATHAN KRAUSE
King's College London

Routledge
Taylor & Francis Group
LONDON AND NEW YORK

First published 2013 by Ashgate Publishing

Published 2016 by Routledge
2 Park Square, Milton Park, Abingdon, Oxfordshire OX14 4RN
711 Third Avenue, New York, NY 10017, USA

First issued in paperback 2016

Routledge is an imprint of the Taylor & Francis Group, an informa business

British Library Cataloguing in Publication Data
Krause, Jonathan.
Early trench tactics in the French Army : the Second Battle of Artois, May–June 1915. – (Ashgate studies in First World War history)
1. World War, 1914–1918 – Trench warfare. 2. France. Armée – History – World War, 1914–1918. 3. World War, 1914–1918 – Campaigns – France – Artois.
I. Title II. Series
940.4'24–dc23

Library of Congress Cataloging-in-Publication Data
Krause, Jonathan.
Early trench tactics in the French Army : the second Battle of Artois, May–June 1915 / by Jonathan Krause.
p. cm. – (Ashgate studies in First World War history)
Includes bibliographical references and index.
ISBN 978–1–4094–5500–4 (hbk)
1. World War, 1914–1918 – Trench warfare – France – Artois. 2. World War, 1914–1918 – Campaigns – France – Artois. 3. France. Armée. Division d'infanterie, 34e. 4. France. Armée. Division d'infanterie, 77e. 5. Tactics. 6. Artois (France) – History, Military.
I. Title. II. Title: Second Battle of Artois, May–June 1915.
D545.A75K73 2013
940.4'24–dc23

ISBN 13: 978-1-138-24950-9 (pbk)
ISBN 13: 978-1-4094-5500-4 (hbk)

EARLY TRENCH TACTICS IN THE FRENCH ARMY

In the English-speaking world the First World War is all too often portrayed primarily as a conflict between Britain and Germany. The vast majority of books focus on the Anglo-German struggle, and ignore the dominant part played by the French, who for most of the war provided the bulk of the Allied soldiers fighting on the Western Front. As such, this important and timely book joins the small but growing collection of works offering an overdue assessment of the French contribution to the Great War.

Drawing heavily on French primary sources the book has two main foci: it is both an in-depth battle narrative and analysis, as well as a work on the tactical evolution of the French army in Spring 1915 as it endeavoured aggressively to come to grips with trench warfare. This period is of crucial importance as it was in these months that the French army learned the foundations of trench warfare on which their conduct for the remainder of the war would rest. The work argues that many advanced practices often considered German innovations – such as the rolling barrage, infiltration tactics, and the effective planning and integration of artillery bombardments – can all be traced back to French writing and action in early 1915.

The work argues that – contrary to received opinion – French army bureaucracy proved effective at very quickly taking in, digesting and then disseminating lessons learned at the front and French commanders proved to be both effective and professional. Such radical conclusions demand a fundamental rethink of the way we view operations on the Western Front.

About the author

Jonathan Krause is a Visiting Lecturer in the Defence Studies Department, King's College, London at the Joint Services Command and Staff College, Shrivenham. He studied history at the University of California, Riverside (2002–2006) before earning an MA (2007) and PhD (2010) from the War Studies Department, King's College, London. Jonathan has published in both French and English on a range of topics related to the French Army in La Grande Guerre, and is an Associate Member of the British Commission for Military History.

Ashgate Studies in First World War History

Series Editor

John Bourne
University of Birmingham, UK

The First World War is a subject of perennial interest to historians and is often regarded as a watershed event, marking the end of the nineteenth century and the beginning of the 'modern' industrial world. The sheer scale of the conflict and massive loss of life means that it is constantly being assessed and reassessed to examine its lasting military, political, sociological, industrial, cultural and economic impact. Reflecting the latest international scholarly research, the Ashgate Studies in First World War History series provides a unique platform for the publication of monographs on all aspects of the Great War. Whilst the main thrust of the series is on the military aspects of the conflict, other related areas (including cultural, political and social) are also addressed. Books published are aimed primarily at a post-graduate academic audience, furthering exciting recent interpretations of the war, whilst still being accessible enough to appeal to a wider audience of educated lay readers.

Also in this series

Britain, Russia and the Road to the First World War
The Fateful Embassy of Count Aleksandr Benckendorff (1903–16)
Marina Soroka

British Generalship during the Great War
The Military Career of Sir Henry Horne (1861–1929)
Simon Robbins

The Ordeal of Peace: Demobilization and the Urban Experience in Britain and Germany, 1917–1921
Adam R. Seipp

To my parents,
all of you,
for your love and support.

Contents

Series Editor's Foreword

France's Great War was an epic of national resistance and organisation. Despite a considerably smaller population than her main enemy, Germany, and her main ally, Great Britain, France and her empire mobilized 8.4M men, suffering 1.3M battle deaths, 4.2M wounded and 537,000 prisoners of war and missing, a total loss (in relation to those mobilized) of 73% (the equivalent German figure was 66% and the British 36%). After the almost disastrous opening battles of the war on the Western Front, France was also compelled to conduct the war with a German army occupying a tenth of French soil, including many centres of industrial production and mineral wealth, only sixty miles from Paris, sometimes closer. In the west, until the British armies arrived in force and took on a more significant part in the fighting, with the opening of the Somme campaign in July 1916, the French army was principally responsible for taking the fight to the enemy. At no time was this more true than in 1915. Gary Sheffield has characterized 1915 as 'the black hole of First World War military historiography'. The French army's battles in the spring and autumn of 1915 were on a massively greater scale than those of the British, but there has been a dearth of studies, especially in English. Jonathan Krause's book is an important addition to our knowledge of a key year of the war and of the leadership and tactical evolution of the French army. It is one of the first fruits of the work being done at Kings College, London, by Professor Bill Philpott and a circle of young scholars aimed at putting the French Army back at the centre of our understanding of the military conduct of the First World War and is therefore doubly welcome.

John Bourne
Centre for First World War Studies
University of Birmingham
October 2012

List of Figures and Tables

Figures

Tables

Acknowledgements

This book started its life as my PhD thesis. It took me three years to turn a desire to better understand the French army into a working piece of scholarship. From then it would take a further year to get the book into its current shape. There have been countless individuals who have helped me along the way, knowingly or unknowingly; thanking them all by name would be impossible. Hopefully those who go unnamed will see some small part of themselves here or elsewhere in the book and know that my gratitude and thoughts go out to them. I would start by thanking King's College, London for the studentship which paid for my doctoral research, and the Society for the Study of French History, which supplied me with a generous grant with which I conducted the last and longest of my research trips to the archives in Paris. I could never have reached this point without access to the extensive holdings at the British Library, the National Archives, the Liddell Hart Centre for Military Archives and the Institute of Historical Research (IHR) among others. Being a history of the French army this book has profited immensely from the *Service Historique de la Défense* (SHD) at Vincennes. The hard-working archivists of the SHD have been a constant help in my years of research. Their prompt assistance and advice were invaluable, even if they could not always do magic. Laurent Henninger of the *Centre des Hautes* Études *Militaires* was always there to provide a stimulating question or thought. I also want to thank the Laubie family, and especially Valérie, for taking me into their home on my many research visits to Paris. Their friendship, support and help with the French language have been priceless: À raison de *vous je sais qu'un canard ce n'est pas "un avion qui nage" et que je ne dois jamais utiliser la phrase "je suis excité" à décrire mes sentiments avant une visite à la marché locale!*

I cannot thank William Philpott deeply enough. A constant guide and mentor for the past seven years, he has been an unflagging supporter and source of inspiration. Without him I might have foundered and lost my way, might have lost focus or drive, might have been unable to come to study in the UK at all, and might have been stranded in Paris without money one particular archive trip. Simply put, without Bill I would not have become a historian of France. To John Bourne I also owe a debt. Years before he was to be my editor (efforts for which I am thoroughly grateful) he played a key role in my early steps as a First World War historian. His influence on me is printed here before you in the form and style of my writing, which he has helped mould. Especial thanks to my friend and mentor Christopher Schmitt for always subtly challenging me to be better, brighter and stranger; I'll see you at the next show.

Friends and colleagues have all played an essential role in facilitating my research. The History of Warfare crew have provided a lifetime's worth of insight, stimulation and friendship. The members' intelligence, expertise, ambition and honesty have sustained me for years, and will undoubtedly continue to do so. Members of the crew will revolutionize their chosen fields, and I am honoured to count myself in their number. The First World War Operations Research Group (the 'Coal Hole Club') at King's has been an important forum for me to express my ideas and receive expert feedback from researchers on the cutting edge. Likewise the regular attendees of the Military History seminar at the IHR, among whom I probably should not count myself, have always been a valuable source of feedback and novel interpretations. Shaunna, I always relied on your support, humour, and unparalleled understanding. Katie Welch, what can I say besides 'eight days a week' …

Of course, I cannot overlook my family. My parents have been an irreplaceable font of love and support without which I never could have survived; it is hard to live so far away from them. Candice, Megan and Katie, you are the most beautiful things in the world to me. Thank you for all of your love, our memories and shared smiles. Adam and Kyle, I am indeed your brother and look forward to sharing life with the young men you are fast becoming. To Lena and Charles, who let me live with them for three years while I did the research that led to this volume: thank you so much for everything; I can never repay your generosity. Thank you both for giving me my love of Wales and of Welsh rugby. *Cymru am byth!* Thank you to Sonya and Brian for being wonderful grandparents. Grandma Judi and Aunt Tanya, you know how much I love you. My heart will forever be with you on a hot summer's day in California: Trinity Lake, Mann's Hole, dusty mountain roads and an old pickup truck. Lastly, to my Papa and Tia: I love you best of all.

> I've been in a worse hole scores of times. It's only a question of persisting. Remember Foch's maxim: '*Attaquez! Attaquez! Attaquez!*'
>
> 'Boris' from George Orwell's *Down and Out in Paris and London*

Czech Memorial This memorial to Czech soldiers in the Foreign Legion stands beside the Neuville – Souchez road, which the Moroccan Division crossed on 9 May 1915. The caption reads: 'They chose to die for liberty'.

Notre Dame de Lorette The Chapel of Notre Dame de Lorette. Housed here are the remains of 43,000 fallen soldiers from the Great War.

General Barbot This monument to the 77e DI stands beside the main road which runs through Souchez. Barbot is depicted in his greatcoat and beret, boldly leading his alpins and chasseurs. The monument is not far from where Barbot was killed.

General Maistre Atop Notre Dame de Lorette is a monument to General Maistre, commander of XXI CA. The CA would capture Notre Dame de Lorette during the course of Second Artois.

Polish Memorial Standing opposite the Czech memorial along the Neuville – Souchez road, this memorial commemorates Polish members of the Foreign Legion which went over the top with the Moroccan Division on 9 May 1915. Large numbers of Polish, Czech, Greek and Russian Legionnaires took part in the attack.

Introduction

Military effectiveness on the First World War battlefield has been an important theme of historical research over the last few decades. Allan R. Millett and Williamson Murray's impressive three-volume text (spanning both World Wars and the inter-war period) remains the pre-eminent work on the subject. The first volume, which deals with the First World War, provides a very useful and meticulously laid out framework for studying military effectiveness at four different levels of war: political, strategic, operational, and tactical.[1] Millett and Murray define military effectiveness as 'the process by which armed forces convert resources into fighting power'.[2] This is an important definition for military historians to be conscious of as it, quite correctly, reduces the importance of 'victory' in any scholarly discussion of military effectiveness. It is one of the essential hurdles for students of military history to surpass: the recognition that a particular force can do everything 'right' and still suffer defeat. What should be of importance to the military historian is not so much the outcome (although understanding an outcome is indeed important), but the military processes that shaped the conflict or engagement being studied. Millett and Murray support this sentiment, saying that 'victory is not a characteristic of an organization but rather a result of organized activity. Judgements on effectiveness should retain some sense of proportional cost and organizational process.'[3]

The judging of military systems in terms of how well they apply the force available to them is a particularly valuable exercise when discussing the First World War. Owing to the unique situation on the Western Front, in which large concentrations of soldiers battled over a relatively narrow band of territory, it can be difficult to accurately judge military effectiveness. This is especially true when action on the Western Front is compared to other more traditional and mobile conflicts in history. The result of such comparison is rarely of much historical value. Fortunately, recent years have seen an increase in the number of studies which seek to better appreciate the various logistical and technological constraints under which the various armies on the Western Front had to operate. In Britain, in the 1990s, a new understanding of operations began to emerge along with a new concept: the 'learning curve'. The learning curve has never been formalized into any sort of concrete way of looking at the war; it is instead a general principle put forth by many different authors over a number of years. The main thrust of the argument is that, contrary to popular opinion

1 Millett, Allan R. and Murray, Williamson (1988), *Military Effectiveness, Volume I: The First World War* (London: Unwin Hyman), pp. 2–4.

2 Ibid., p. 2.

3 Ibid., p. 3.

in Britain, the British army did not stagnate and flounder during the war. Rather, Britain emerged from the conflict as a modern fighting force which had effectively tackled problems of command, control and logistics and ultimately went on to play a crucial role in the defeat of the German army in the field in 1918.[4]

Despite the growing awareness of the British army's military methods, there remains a lack of understanding of the French army as a military organization. Douglas Porch's chapter on the French army in the First World War in Millett and Murray's book illustrates many of the commonly held misperceptions of the French army. As if echoing the popular 'mud and blood' portrayal of the First World War that British scholarship has been addressing now for some time, Porch refers to the Western Front as a 'narrow band of congealed horror' and bemoans the 'irrationality' of the conflict.[5] He pointedly asks why no one stopped to consider whether or not the price paid in human lives for victory over Germany was 'too high'. Porch casts down blame, as if from a pulpit, upon generals, politicians and even the common soldiers who, he claims, failed to ask this pertinent question and thus contributed to the mindless slaughter of a futile war.

Nevertheless, Porch is not quite as polemical as some of his less-scholarly analogues who have written on the British military experience of the war.[6] He acknowledges French advances in 'aviation, artillery tactics and motorized transport' and compares the French generalship very favourably to the British, which he unfairly implies was probably the worst on the Western Front.[7] Porch asserts that 'the French soldiers made many miscalculations on the strategic, tactical and technical levels cannot be in doubt. However, to denounce them simply as incompetent, or worse, as men who placed professional ambition above the lives of their troops or the long-term interests of France, leaves too many questions begging.'[8]

Prima facie, these seem to be at least reasonably fair evaluations and comparisons such as one might expect from a modern, even-handed work of scholarship. However, they are quickly belied by Porch's dogged adherence to the assumption of German superiority in all things tactical. Porch claims that 'French commanders, even the best ones, lagged behind their German foe in developing the

[4] Griffith, Paddy (1994), *Battle Tactics on the Western Front: The British Army's Art of Attack, 1916–1918* (London: Yale University Press); Prior, Robin and Wilson, Trevor (1992), *Command on the Western Front: The Military Career of Sir Henry Rawlinson 1914–18* (Oxford: Basil Blackwell); Sheffield, Gary (2001), *Forgotten Victory: The First World War: Myths and Realities* (London: Headline); and Sheffield, Gary and Todman, Dan (2004), *Command and Control on the Western Front: The British Army's Experience 1914–1918* (Staplehurst: Spellmount) are all excellent books in this strain.

[5] Millett and Murray, *Military Effectiveness*, pp. 190–191.

[6] Clark, Alan (1962), *The Donkeys* (London: Hutchinson).

[7] Millett and Murray, *Military Effectiveness*, p. 191.

[8] Ibid.

training, tactics, and staff work … which might have saved many French lives.'[9] He admonishes the 'cavalier disregard for human life sometimes exhibited by the [French] high command' and boldly claims that French 'military thinking hardly rose above the level of the rather forced optimism of popular clichés like, "*On les aura*"'.[10] According to Porch, the French 'seldom exhibited great originality [and were] content to copy German techniques' which favoured a decentralized, dynamic form of command which Porch contrasts sharply with the centralized Allied methodology: this argument appears frequently when discussing the differences between the Germans and the Allies on the Western Front.[11] Porch goes yet further in his uncritical lauding of the efficient tactical German war machine, stating that 'even when French soldiers managed to gain local successes, they were unable to exploit them due to the rigid pre-planning of attacks and to the efficiency of German infantry in the murderous business of trench warfare'.[12] Not only does this assume a sort of inherent German aptitude for trench warfare, but it also neglects the primacy of German artillery in slowing and eventually halting Allied attacks.

Porch repeatedly denigrates the French ability to wage war in any independent fashion by asserting that France should, in the first instance, have waited until Britain had fielded a large army before attacking and then, having failed to do that, should have waited (as Pétain advocated) for 'the tanks and the Americans'.[13] Porch does not make a convincing argument that France was incapable of offensive action; nor does he consider the effect on morale that failing to act would have engendered. Likewise, he ignores the potential impact of failing to lead on France's standing among its allies (a crucial aspect of the political dimension of the war). Porch does not explore in any way the fact that, as the British showed on 1 July 1916, having large quantities of men and guns was, in and of itself, insufficient. Experience in trench warfare, and in the administration of large attacking forces, was essential to success on the Western Front; such experience could not be gained by idle inactivity. Tactically, Porch claims that while French artillery plans got longer and more resource-intensive they failed to be innovative in their approach (despite having earlier listed 'artillery tactics' as a specific success of the French in the First World War). He goes on to claim that 'until the summer of 1917, France made war essentially with its infantry' who were 'armed with little more than personal courage', which not only contradicts his repeated attacks on the extended French preparatory bombardments, but also fails to recognize the radical transformation that the French military machine underwent over the course of the war in which

[9] Ibid.

[10] Ibid., pp. 191 and 206.

[11] Ibid., p. 211 and Gudmundsson, Bruce I. (1989), *Stormtroop Tactics: Innovation in the German Army, 1914–1918* (Westport, CT: Praeger), p. 173.

[12] Millett and Murray, *Military Effectiveness*, p. 215.

[13] Ibid., pp. 204 and 207.

grenades, gas, flame-throwers, light machine guns, combat aircraft, bombers, and tanks all became integrated into a highly complex, modern combat system.[14]

Fortunately, Porch's vision of the French army in the First World War is beginning to be challenged, as evidenced by works such as Michel Goya's *La chair et l'acier: l'invention de la guerre moderne, 1914–1918*. This book, published in 2004, argues that through its long process of innovation and apprenticeship in the trenches the French army had become Europe's first 'modern' army by the end of the war.[15] It is in many ways the purpose of this book to expand upon the scholarship of Goya and others by providing a detailed operational study of a key moment in the French army's development. This case study of one particular battle which took place in spring 1915, the Second Battle of Artois, will demonstrate how the French addressed the many tactical and operational problems that trench warfare presented, arguing that this early experience provided the foundation of modern tactical and operational methods which enabled them to master their enemy by late 1918.

There are events in the Great War which are considered to be truly defining moments of the conflict on the Western Front: the Miracle of the Marne; 1 July 1916; Verdun; 8 August 1918, the 'Black Day' for the German army. These events were not only important for their military results, but also for their psychological impact on the course of the war and how we remember it. Had small things gone differently, Second Artois (9 May–18 June 1915) might well have been included on the same list as those revered coincidences of time and space. It was the largest of the early trench battles, and saw a greater commitment of French infantry than was made at the start of the Battle of the Somme.[16] Second Artois furthermore marked a crucial turning point in the conduct of operations on the Western Front, deeply influencing Allied offensive and German defensive tactics. As an example: among many other important developments the French offensive during Second Artois was the first to use 'infiltration tactics'[17] on a large scale, an innovation usually attributed to German stormtroopers much later in the war.[18]

The French successes on 9 May, feats aided by their doctrinal overhaul before the battle, contributed heavily in compelling the German High Command to radically alter their defensive tactics. Their thin, improvised trench network would be quickly strengthened in order to ensure that Germany could continue

---

14 Ibid., pp. 207 and 212.

15 Goya, Michel (2004), *La chair et l'acier: l'invention de la guerre moderne, 1914–1918* (Paris: Tallandier).

16 Doughty, Robert (2005), *Pyrrhic Victory: French Strategy and Operations in the Great War* (Cambridge, MA: Belknap Press of Harvard University Press), p. 291.

17 As a definition for infiltration tactics I focus on its main tenets of a violent forward thrust which targeted weak points in the enemy defences while bypassing enemy strong-points, which would in turn be dealt with by reserves following closely behind the initial attack.

18 Gudmundsson, *Stormtroop Tactics*.

to prioritize manpower and materiel to efforts on the Eastern Front. The German army would counter the newfound French prowess by digging secondary positions sited on reverse slopes and deep underground bunkers for protection against French artillery bombardments. Pre-sited artillery barrages would be coordinated to ensure a rapid response to any developing attack. The Germans would also begin to array their defence in depth, using thinly manned lines of machine-gunners to break the cohesion of Allied attacks. This transformation of the German defensive network gave birth to the trench-warfare paradigm so widely caricatured in popular history: the Germans hiding in deep underground bunkers for days on end, enduring Allied bombardments, only to emerge unscathed and ready to pour deadly machine-gun fire into dense masses of advancing French or British infantry. The German emphasis on defence on the Western Front should call into question the perception held by some scholars who prefer to paint the Germans as the pre-eminent innovators of offensive tactics.[19]

The influence of Second Artois was not limited to tactics, but was also a major impetus for operational, strategic and doctrinal evolution in the French army. It was often cited as proof that the Allies could 'break through' the enemy defences and end the trench stalemate on the Western Front. Second Artois made the need for modern heavy artillery even more pressing and apparent, while at the same time yielding valuable experience to French commanders in the organization and execution of large-scale trench offensives. Second Artois was influential not merely over the course of 1915, but also played an important role in the evolution of the entire conflict. The tactical doctrine on which Second Artois was based, «But et conditions *d'une action offensive d'ensemble*» (also known as Note 5779), contained early drafts of key concepts, many of which would remain important to the continued doctrinal evolution of the French army in the First World War. The document would undergo two revisions during Second Artois, and many of the key ideas that Second Artois helped solidify remain clearly present in doctrine written by General Ferdinand Foch in 1916 and employed during the Somme and in later French offensives.

At 10.00 on 9 May 1915 the French Tenth Army, commanded by General Victor d'Urbal, launched an attack along a 19-kilometre front, and made one of the most impressive advances of the First World War. On this day the XXXIII *Corps d'Armée* (CA), commanded by General Philippe Pétain, advanced 4.5 kilometres in a mere hour and a half, storming Vimy Ridge and investing a handful of villages and woods. By noon advance guards of the *Division Marocaine* (DM) and the 77e *Division d'Infanterie* (DI) were engaging the Germans as far east as Givenchy-en-Gohelle, just behind the northern shoulder of Vimy Ridge.[20] The advance was so spectacular that staff officers in Tenth Army refused to believe it for several hours,

19 Griffith, *Battle Tactics on the Western Front*, p. 25 and Gudmundsson, *Stormtroop Tactics*, p. 172.

20 There remains to this day a memorial atop Vimy Ridge commemorating the DM's storming of the hill.

and held back crucially needed reserves. This unprecedented success was made possible by an incredibly detailed and well-executed artillery preparation, as well as the successful prosecution of early infiltration tactics by the advancing infantry. Both can be attributed to the diligence of General Pétain, who had taken such a keen and personal interest in the operation of his batteries that Basil Liddell Hart would go on to joke in his biography of Pétain that he had taken the time to aim every gun himself.[21]

The advance of the XXXIII CA, which inaugurated a new approach to trench warfare, was ultimately unsustainable. Reserve troops found themselves up to 8 kilometres from the fighting by the time they were ordered to move forward and support the assault waves. As a result, the first reinforcements (the 67ᵉ Brigade) did not reach the French starting position until noon, two hours after the initial attack had been launched. 18ᵉ DI, sent from army reserve, would not reach that point until 17.15, three hours after a German counter-attack pushed the DM and 77ᵉ DI out of Givenchy and forced them to dig in in an untenable defensive position before *cote* 140 (the designation given to Vimy Ridge). Reinforcements were continually fed into the maelstrom at Vimy for the next three days until, under intense and insurmountable pressure, the battered units of the DM and 77ᵉ DI were forced to admit their inability to capture and hold the ridge. They would be the last Allied troops to set foot on top of Vimy until the Canadians finally captured that dominating feature two years later.

If the advance made on 9 May could not be consolidated, its psychological impact upon French command and operations would be more permanent. In his 1925 book examining tactical evolution in France and Germany during the war André Lucas points out that 'the offensive in Artois showed that the rupture of a solidly organized front that was not comprised of only one single position of resistance was perfectly realisable, on the condition that one had the necessary amount of heavy artillery'.[22] This lesson was not merely post-war hindsight but was found immediately attractive to the high command, especially Joffre, who felt that the breakthrough would in fact have occurred if Tenth Army had had more guns and more shells, and if the British forces on the French left flank had not performed as poorly as they had, thus allowing the Germans to concentrate their reserves against the French push on *cote* 140.[23]

---

21 Liddell Hart, Basil (1937), *Foch: The Man of Orleans* (London: Penguin), p. 192.

22 Lucas, Pascal Marie Henri (1925), *L'Evolution des idées tactiques en France et en Allemagne pendant la guerre de 1914–1918* (Paris: Berger-Levrault, 1925, third edition), p. 67: «*L'offensive d'Artois a montré que la rupture d'un front solidement organisée et ne comportant qu'une seul position de résistance est parfaitement réalisable, à condition de disposer du matériel d'artillerie lourde nécessaire.*»

23 Daille, Général M. (1936), *Histoire de la Guerre Mondiale: Joffre et la guerre d'usure 1915–1916* (Paris: Payot).

This continued belief in the *percée*[24] would dominate planning for the unsuccessful offensives in Champagne and Artois in September 1915. Even Pétain, contrary to popular belief in his unflagging caution, was convinced that the *percée* was an achievable goal, stating in his after-action report: 'The attack of the 9th proves that a breakthrough is possible.'[25] After this assertion Pétain offered up what he considered to be the appropriate steps necessary to ensure an adequate preparation for an offensive:

> [An attack] ought to be methodically and minutely prepared (terrain organized, *parallèles de départ*, communication trenches [*boyaux*] and assembly points created, research done on objectives [from the enemy's first line] until their last and the simultaneous preparation of all of these objectives by artillery) …
>
> [An attack] will move very quickly when it is successful (the combat at NEUVE CHAPELLE has already proven this). As the breach made will close as quickly as it was opened, if the enemy has reserves on hand we have come to the conclusion that to succeed it is necessary to sustain the attack with exploitation troops which follow immediately behind the attacking troops …
>
> [The attack] is easier in open terrain than it is in terrain with cover. Strong-points which are made up of villages are all particularly difficult to capture …
>
> [An attack] ought to be made on a considerable front. The enemy in this case cannot effect enfilade artillery fire on the assailants; strong-points which do not fall [initially] can be easily outflanked; the strategic result of manoeuvre could become considerable.[26]

---

24 The French term for 'breakthrough'.

25 *Service Historique de la Défense* (*SHD*), 22N1832; «*Compte-rendu Sommaire des Opérations du 33e Corps d'Armée pendant les journées des 9, 10 et 11 Mai 1915*»: «*L'attaque du 9 prouve que la percée de l'ennemi est possible.*»

26 Ibid. «Elle *doit* être *méthodiquement et minutieusement préparée (organisation du terrain, création de parallèles de départ, de boyaux de communication, de places de rassemblement; recherche des objectifs jusqu'à la dernier ligne ennemi, et préparation simultanée d'artillerie sur tous ces objectifs); … Elle s'effectue très rapidement quand elle doit réussir (le combat de NEUVE CHAPELLE l'avait* déjà *prouve). Comme la brèche se referme* également *très vite si l'ennemi dispose de quelques réserves, on en arrive à cette conclusion qu'il faut pour réussir amener d'avance les grosses unités chargées d'exploiter le succès au contact immédiat des C.A. d'attaque; … Elle est plus facile en terrain libre qu'en terrain couvert. Les points d'appui constitués par les villages sont tout particulièrement difficiles à enlever; … Elle doit* être *faite sur une front considérable. L'ennemi dans ce cas ne peut en effet croiser les feux de son artillerie sur les assaillants; les points d'appui qui ne tombent pas sont facilement tournés; le résultat stratégique de la manœuvre peut devenir considérable.*»

The above was sent to *Grand Quartier Général* (GQG) and would mirror the 'official' lessons learned from Second Artois. These lessons were very clearly evolved from Note 5779, and would be carried on to become the basis for the September 1915 offensives, albeit in slightly modified form. This begs the counter-factual query: had the XXXIII CA not achieved such stunning success in May 1915, would the French have stopped chasing the elusive *percée*? If the French had reassessed their operational planning in spring the battles of autumn would have been radically different, and may have had much smaller, more realizable objectives. This in turn would likely have influenced the planning and execution of subsequent battles (Verdun and the Somme especially). In terms of physical capability one can venture to say that a more conservative and well-planned offensive on 25 September 1915[27] might have cost far fewer French casualties in men and materiel, making the French much stronger during the 1916 battles – which might have made them even more successful for the Allies[28] and damaging for the Germans than in fact they were. While it is easy to get carried away with counter-factual history it can help illuminate the crucial importance of certain events in history, such as the negative psychological and doctrinal effect of the XXXIII CA's success on May 9.

Of course not everyone was convinced that the success in Second Artois could be repeated – indeed not only repeated but also enhanced – and achieve a breakthrough. The most senior and vocal opponent to a continued quest for the *percée* was Ferdinand Foch, who commanded the *Groupe Provisoire du Nord* (GPN),[29] of which d'Urbal's Tenth Army was a part. After Second Artois had ended Foch asserted that: 'In the current circumstances, before an enemy established for the long haul, it appears wise to not place all of our hopes, nor risk all of our forces, on the possibility of a *percée*, on the idea of a victorious and decisive breakthrough by numerical superiority … the *percée* remains doubtful.'[30] At several different junctures in spring 1915 and beyond Foch would step in and repeat his mantra of capturing small, tactically important pieces of terrain step by step by advancing 'surely, methodically and economically'.[31]

---

27 The largest single day of offensive action in 1915: the French launched two major battles (Second Champagne and Third Artois) in conjunction with the British attack at Loos.

28 The idea of 1916 and the Somme in particular as being successful for the Allies is certainly contentious and is deserving of a much more in-depth discussion than could possibly be given here. For more see Philpott, William (2009), *Bloody Victory: The Sacrifice on the Somme and the Making of the Twentieth Century* (London: Little, Brown), among others.

29 Later the Groupe des Armées du Nord

30 Daille, *Histoire de la Guerre Mondiale*, p. 127: «*Dans les circonstances habituelles, devant une* établissement *de longue durée de l'ennemi, il paraît sage de ne pas fonder toutes nos espérances, ni risquer toutes nos forces disponibles, sur la possibilité de percer, sur l'idée d'une trouée victorieuse et décisive, par la puissance de nombre … la percée reste douteuse.*»

31 Marshall-Cornwall, General Sir James (1972), *Foch as Military Commander* (London: Willmer Brothers), p. 178.

General Fayolle, commanding 70ᵉ DI in Pétain's XXXIII CA, was also opposed to any sort of *percée* doctrine and preferred a slow and methodical approach. When commenting on those who favoured a 'continuous battle' (a concept which sought to create a breakthrough by constantly feeding in fresh waves of infantry) Fayolle remarked in his published diary that 'their illusion is grand'.[32] Despite their later friendship and mutual respect he was strongly at odds with Foch during this time, blaming Foch for ordering his subordinate commanders to undertake tasks beyond their physical means to do so. Fayolle implicated not only Foch but also d'Urbal in this vein, saying 'Foch and d'Urbal have not taken account of the state of the battlefield or of the necessities of trench warfare.'[33] Fayolle was not alone in thinking that the higher echelons were oblivious to the conditions at the front. One young artilleryman who fought in Second Artois, André Laffargue, was so horrified at what historian General Daille called 'flagrant tactical failures' that he decided to produce a pamphlet advocating for change in the French army's tactical/operational methodology.[34]

That pamphlet, Étude *sur l'attaque dans le période actuelle de la guerre*, was published in mid–late 1915 and would eventually be widely distributed through both the French and the British armies. It was perhaps distributed too widely; it was quickly captured and studied by the Germans. Its primary contribution to French thinking was an increased reliance on firepower (especially heavy artillery/mortars), and a much more cautious approach to trench warfare which would become the standard by 1916–17. Before overstating the importance of Laffargue and his pamphlet it should be said that historians are still unsure how much influence Laffargue's writing exerted on the thinking of officers in the field, and subsequently on how these officers conducted operations.[35] It is perhaps best to consider Laffargue as evidence of the tactical evolution that the French army was experiencing so sharply in 1915 with the shift from an infantry-intensive to an artillery-intensive battle.

Needless to say, operational and tactical evolution was not the sole prerogative of the French, but was being undertaken by all the armies on the Western Front. In his study of the development of battlefield tactics André Lucas asserts that it was the French success on 9 May that convinced the Germans of the necessity to dig stronger trench networks comprising several different positions arrayed in depth. This concept has been supported by other historians, including Robert T. Foley and

---

[32] Fayolle, Marie Émile (1964), *Cahiers secrets de la Grande Guerre* (Paris: Librairie Plon), p. 97: «*Leur illusion est grand*».

[33] Ibid., p. 113: «Foch et *d'Urbal ne se rendent aucun compte de l'état du champ de bataille et des nécessités de la guerre de tranchées.*»

[34] Daille, *Histoire de la Guerre Mondiale*, p. 123.

[35] Goya, *La chair et l'acier*, p. 206; Griffith, *Battle Tactics on the Western Front*, p. 56; and Gudmundsson, *Stormtroop Tactics*, p. 173.

Jean-Claude Laparra.[36] This concept was put down in a memorandum in June of 1915 that stated that by having two or three lines of defence 1–2 kilometres behind each other it would necessarily force the attacking enemy to move their artillery forward before attacking the second or third German positions. This would buy the defenders enough time to funnel reserves into the *schwerpunkt*[37] and thereby prevent a disaster from developing.[38] On 7 June 1915 XI Bavarian Corps, stationed in the Artois sector, captured a copy of Note 5779, which laid out the French way of warfare based on an intense artillery bombardment followed immediately by rapid penetration of the German position. In response, XI Bavarian Corps sent an internal note to its subordinate commanders highlighting the need for intensive digging to begin immediately in an attempt to strengthen their local defences. The Bavarian corps commander demanded, profoundly, that the second position be every bit as solid as the first.[39]

The above was to have significant implications for the remainder of the war. Across the Western Front German defensive positions were soon made truly formidable. The newly dug second positions were indeed well behind the first line and often sited on the reverse slopes of hills, rendering them invisible to terrestrial observation and increasingly difficult for French artillery to hit. Such a disposition realized Hippolyte Langlois' pre-war fear of trench fortifications arrayed in depth, which he theorized would be 'invulnerable to artillery'.[40] Speaking from experience, Fayolle commented that 'as for taking many successive lines, those which we cannot see will be intact'.[41] This is in fact exactly what befell many attacking *poilus* on 25 September 1915 as wave after wave fell before the unbroken wire of the German second position, which had received virtually no bombardment, especially in the Fourth Army sector in Champagne.

The solidification of the German trenches as a result of their experience in Artois implies some fault with their trenches early in 1915. Already one rapid advance had been made against the weak German trenches north of Arras during

---

36 Lucas, *L'Evolution des idées tactiques en France et en Allemagne pendant la guerre de 1914–1918*, p. 75; Foley, Robert T. (2005), *German Strategy and the Path to Verdun: Erich von Falkenhayn and the Development of Attrition, 1870–1916* (Cambridge: Cambridge University Press), p. 163; Laparra, Jean-Claude (2006), *La Machine à Vaincre, De l'espoir à la désillusion: Histoire de l'armée allemande 1914–1918* (Mercuès : Imprimerie France Quercy), p. 121; and Sheldon, Jack (2005), *The German Army on the Somme 1914–1916* (Barnsley: Pen and Sword Books), p. 66.

37 The 'decisive point'

38 Lucas, *L'Evolution des idées tactiques en France et en Allemagne pendant la guerre de 1914–1918*, p. 76.

39 Ibid.

40 Langlois, Hippolyte (1908), *L'Artillerie de campagne en liaison avec les autres armes* (Paris: Librairie Militaire R. Chapelot), p. 138.

41 Fayolle, *Cahiers secrets de la Grande Guerre*, p. 98: «Quant à *emporter les lignes successives, dont plusieurs, celles qu'on ne voit pas, seront intactes*.»

the British attack on Neuve Chapelle.[42] It is interesting to note that Pétain draws direct reference to this engagement when putting forth his belief in the *percée* in the after-action report quoted above.[43] Foley confirms the poor quality of the German trenches during this period: 'In most cases, these systems reflected the pre-war emphasis on temporary field fortifications. Many positions lacked depth and were consequently vulnerable to powerful enemy attacks.'[44] Foley goes on to emphasize that these problems would only just start to be addressed by spring–summer 1915, which is just when Second Artois was raging.

Given the relative weakness of the German defences, and the recent transfer of 10½ German divisions from the Western to the Eastern Front, the Allies had an opportunity to achieve an important victory.[45] Was a breakthrough really possible? Probably not; but even if the French could not have broken through their advance might still have yielded Vimy Ridge before the Germans could respond. As the eastern side of Vimy Ridge is almost sheer, it would have been very difficult for the Germans to attempt to retake it. Having a nearly impenetrable bastion with a commanding view of the plains below in their hands the French could have pursued operations in the region with relative ease. French artillery deployed on the reverse slope of Vimy Ridge could fire with impunity across most of the Douai plains and propel the French forward towards major railway junctions such as Lille. The capture of such a critical junction would have had very serious implications for German operations in France, perhaps even prompting a retreat similar to the one after the Marne campaign in 1914. Refraining from further conjecture, this work would reassert that Second Artois was a seminal turning point in the early phases of trench warfare and deserves recognition and scholarly attention for this fact.

Historiography and Sources

Historians have tended to overlook the French in the First World War; Anglophones typically only discuss the French when doing so enriches the British or American narratives. What coverage there is tends to focus on certain well-known events, including the 1917 mutinies and the disastrous French '*offensives à outrance*' in 1914. When 1915 is taken into consideration it is too often summarized as a pointless year of 'mud and blood'. This is perhaps an understandable pitfall considering the fact that nearly 350,000 *poilus* died that year – more than in any

---

42 Edmonds, Brigadier-General Sir James E. (1928), *Military Operations in France and Belgium, 1915* (London: Macmillan), p. 12.

43 *SHD*, 22N1832; «Compte-rendu *Sommaire des Opérations du 33e Corps d'Armée pendant les journées des 9, 10 et 11 Mai 1915*».

44 Foley, *German Strategy and the Path to Verdun*, p. 163.

45 Daille, *Histoire de la Guerre Mondiale*, p. 110.

other year during the Great War.[46] For the 50th anniversary of 1915 the *Revue Historique des Armées* dedicated an entire issue to the study of this year. Despite the fact that this volume is now over 40 years old it remains a fairly accurate barometer of the state of scholarship on 1915.

In the introduction to this particular volume Maurice Genevoix states that:

> When considering the relations of historians of the First World War [to 1915] one cannot help but be struck by the evident lack of interest which they exhibit towards this period … 1914: the drama of the war of movement, the German advance, the 'Miracle of the Marne'; 1916: the battle of Verdun, the heroic and decisive French resistance, *'on les aura'* … Between the two: digging-in, stagnation in the trenches, a few vain 'partial offensives', practically nothing.[47]

In addition to the lack of interest that 1915 typically inspires, the sheer volume of casualties is an obstacle to a better understanding of its many complexities. Genevoix echoes this sentiment:

> At first sight, the facts we have just recounted lead us to an abrupt and summary logic which would seem to yield a terrible formula: 'in 1915, three hundred fifty thousand Frenchmen were made to die for nothing'. Even if this is true it is too simplistic and, as a consequence, unjust.[48]

Genevoix goes on to caution historians against being too harsh on commanders for not adopting the principle of 'the artillery conquers, the infantry occupies' quickly enough, while again reiterating his assertion that it can be very difficult to lift one's focus from the horror of the trenches and see the bigger strategic picture. In the hopes of promoting a more balanced appreciation of the difficult position France found itself in, he cites the contradictory forces pulling the French army in 1915: on one hand they must attack to support Russia and drive the Germans from French soil, while on the other hand they do not yet have the training or the

---

46 Genevoix, Maurice (1965), '1915 "Année Terrible"', *Revue Historique des Armées*, vol. 21, no. 2, p. 5.

47 Ibid. «*Se reporter aux relations des historiens de la Première Guerre Mondiale, on ne manquera pas d'être frappé par l'évidente absence d'intérêt dont elles témoignent envers cette période … . 1914: drame de la guerre de mouvement, la ruée allemande, le 'miracle de la Marne';* 1916*: la bataille de Verdun, l'héroïque et décisive* résistance *française, 'on les aura' … Entre les deux, enfouissement, stagnation dans les tranchées, quelque vaines 'offensives partielles', à peu près rien.*»

48 Ibid. «*Ainsi à première vue, à partir des données que nous venons de rappeler, une logique abrupte et sommaire aboutirait une formule terrible: 'On a fait tuer, en 1915, trois cent cinquante mille Français pour rien.' Même si c'est vrai, c'est trop simpliste et c'est par conséquent injuste.*»

modern heavy artillery with which to do so.[49] Genevoix's eloquent and stirring sentiment represents very forward thinking in relation to the historiography of 1915; unfortunately, not every author represented in the volume followed suit.

Despite Genevoix's warning against an overemphasis on the mud and blood of 1915 the only operational study of the volume chooses to focus on just that. The article on the battle of Les Eparges, written by Lieutenant-Colonel de Chasteigner, is primarily made up of quotes from soldiers' diaries recounting the horrible conditions they faced. Despite doing an adequate job of underscoring the contradictory obligations of the French army in 1915 just as Genevoix had done, de Chasteigner's article exhibits many pitfalls that remain common to the present day.[50] First and foremost, de Chasteigner chooses one of the smallest and least significant battles of the entire year to use as his case study: Les Eparges. This same methodology has been used by historians across the spectrum. John Mosier focused almost exclusively on Les Eparges in his popular *Myth of the Great War*, a battle which he called 'arguably … the most important of the early French offensive efforts'.[51] In a year of four major and several mid-sized battles, choosing such a small engagement as the exemplar for 1915 is doing a gross disservice to the complexity, scale and scope of that formative year. De Chasteigner also refers to the French 'adapting to the style of the adversary', which implies the current popular view of German innovation and efficiency while the Allies are depicted as forever trying to catch up.[52] Ultimately, the *Revue* does not offer the modern scholar much to work with.

There exist only a few major works which directly address Second Artois. These include three English-language sources (Nick Lloyd's *Loos 1915*, Robert Doughty's *Pyrrhic Victory* and Anthony Clayton's *Paths of Glory*), while in French General Palat's *La grande guerre sur le front occidental*, General Daille's *Histoire de la Guerre Mondiale* and the French Official History, *Les Armées Françaises dans la Grande Guerre*, are the main secondary sources. Doughty and Clayton both cover the French war effort in its entirety, a very difficult thing to do in one volume. Only a small amount of space can be allocated to the Artois offensives. Both agree with the popular theory (concurrently held by the three aforementioned French sources) that had Pétain been able to quickly reinforce the 77e and Moroccan divisions on 9 May the French could have held Vimy Ridge.[53] While the 4 1/2 kilometre advance of two-thirds of Pétain's XXXIII CA

---

49 Ibid., pp. 5–6.

50 de Chasteigner, Lieutenant-Colonel (1965), '1915: Le martyre de l'infanterie. un exemple: Les Eparges', *Revue Historique des Armées*, vol. 21, no. 2, p. 8.

51 Mosier, John (2001), *The Myth of the Great War: A New Military History of World War I* (London: Profile Books), p. 135.

52 de Chasteigner, 'Le martyre de l'infanterie', p. 7: «*notre Commandement veut s'adapter au nouveau style de l'adversaire*».

53 Clayton, Anthony (2003), *Paths of Glory: The French Army, 1914–1918* (London: Cassell Military) p. 70 and Doughty, *Pyrrhic Victory*, p. 159.

is remarkable compared to the vast majority of other attacks in 1915, this very brilliance may well have blinded commentators to its full significance in ushering in a new paradigm of trench warfare on the Western Front.

Anthony Clayton wrote that 'the story of 1915, then, is one of a series of fruitless attacks by both the French and the British that, at best and at heavy cost, only moved the opposing trench systems forward for short distances.'[54] Clayton not only makes a common analytical mistake by trying to measure success in trench warfare in terms of ground captured, he also manages to condemn an entire year's worth of effort in a single throwaway sentence. His direct discussion of Second Artois is hardly more engaging. Repeating himself, he asserts that 'the end result differed little from the great August 1914 offensives – an advance of a mere 4 km at the cost of 300,000 casualties, of whom 100,000 were men killed'.[55] No serious consideration of the battle, its strategic implications or logistical and temporal limitations is to be found. His casualty data is also dubious and seems to be drawn, uncritically, from Palat. All other estimates yield figures of between 100,000 and 121,000 total casualties for the French and 50,000–80,000 for the Germans.[56]

Robert Doughty says more than Clayton in his discussion of Second Artois, but still does not sufficiently explain the battle and its full repercussions, owing to time and space limitations. Doughty does a good job of briefly covering both the May and June periods of Second Artois as well as spending a significant amount of time discussing the strategic situation at the time.[57] Crucially, he spends time in *Pyrrhic Victory* discussing the various operational models being suggested in 1915 and accurately contrasts the various competing theories on the appropriate way for the French army to wage war which existed at the time.[58] Nevertheless, Doughty fails to consider much of the tactical evolution present in Second Artois and, despite his account's value, it remains in need of substantial expansion if Second Artois is to occupy its proper place in the historiography.

Most recent of the three important English-language volumes which deal with the war is Nick Lloyd's *Loos 1915*, which covers Second Artois much more quickly than Doughty, and merely so as to provide a context for the main thrust of his work which is, unsurprisingly, the Battle of Loos. Lloyd recognizes the 'tantalizing success … around Souchez and Vimy Ridge' on 9 May, but comes to the same conclusion: that the exposed elements of XXXIII CA, which had nearly captured those two important pieces of terrain, could not hold out in the

54 Clayton, *Paths of Glory*, p. 60.

55 Clayton, *Paths of Glory*, p. 66.

56 Chatinières, Paul, Janet, Paul, Lefranc, Abel and Landousie (1923), *Les armées françaises dans la Grande Guerre* (Paris: Imprimerie Nationale), pp. 101–102 [henceforth *AFGG*]; Doughty, *Pyrrhic Victory*, p. 65; and Palat, Barthèlemy Edmond (1927), *La grande guerre sur le front occidental* (Paris: Chapelot), p. 365.

57 Doughty, *Pyrrhic Victory*, pp. 159–165.

58 Ibid., p. 156.

face of withering enemy fire and the unavailability of reinforcements.[59] Not being an expert on the French army, Lloyd's condemnation of Foch as being out of touch and an adherent of the cult of the offensive as late as 1915 comes as no surprise; this book suggests otherwise.[60] Indeed, the book focuses almost solely on the weaknesses of the French army during this period while mentioning few if any of its strengths, a likely side-effect of Lloyd's use of sources such as Clayton and Douglas Porch, neither of whom maintain much esteem for the French army in the First World War.[61] Notwithstanding this, Lloyd is firmly correct in his view of 1915 as a whole:

> The early trench battles of the war, particularly during 1915, defined what would happen later on as experience was gained, lessons were learnt and soldiers and politicians gradually came to terms with the revolution in warfare that they were witnessing.[62]

Language and cultural barriers being one of the main reasons that Anglophones know so little about the French efforts in the First World War, it should be of no surprise that the best work written on the development of the French army is in French. Michel Goya's PhD thesis, *Le processus d'évolution tactique de l'armée française de 1871 à 1918*, and his book *La chair et l'acier* together cover the French army's tactical and managerial development from 1871 to the end of the First World War. Naturally, these works suffer from the same limitations as Doughty and Clayton's books in that their broad scope leaves very little space for individual engagements. Nevertheless, Goya did find time to address directly Second Artois, and correctly asserts that it was the first battle the French fought that was based entirely on lessons learned from previous trench battles (namely, the first battles of Artois and Champagne).[63] Goya goes on to commend the methodical, scientific preparation that led up to Second Artois, calling it the first '*mathématisation*' of combat, as well as the successes of Pétain's XXXIII CA on 9 May. He is nevertheless critical of GQG and the doctrine of 1915. Goya asserts that the real evolution took place in the form of apprentice learning in the trenches and that doctrinal notes from above were simply discarded by men at the front who knew better. This seems a slightly idealistic view that probably does not reflect the entire complexity of the situation.

Concerning the French attack on 9 May, General Palat highlights how difficult it was for d'Urbal to order any reinforcements forward to assist the 77e and Moroccan divisions as German artillery fire had cut the telephone wires heading

59 Lloyd, Nick (2006), *Loos 1915* (Stroud: Tempus), p. 20.

60 Ibid., p. 25.

61 Ibid., p. 26.

62 Ibid., p. 14.

63 Goya, Michel (2008), *Le processus d'évolution tactique de l'armée française de 1871 à 1918* (doctoral thesis, University of Paris IV), p. 271.

back to d'Urbal's headquarters. Furthermore, when reports of the advance of XXXIII CA reached d'Urbal's staff they refused to believe what they thought to be gross exaggerations.[64] The Official History, in a different strain, emphasizes how General Blondlat, commanding the DM, used all of his reserves to reach the crest of Vimy. This leads one to think that the DM's progress was pushed harder than it need have been, thereby bringing about its untenable conclusion.[65] The Official History is very clear in its assertion that d'Urbal's attempt to push forward the offensive on the days immediately following May 9 was bound to fail, stating that 'all the attacks would fail; our troops [were] tired, the units [were] all mixed together; furthermore, the enemy had brought new reinforcements from Lens and Douai'.[66] This then raises the question of command and generalship in Artois. Were the generals, especially the higher ranking ones, in tune with the realities on the ground as well as with what was necessary to achieve victory, or were they instead trying to fight their own private theoretical battles?

There can be clearly discerned three different operational methods favoured by various leaders during Second Artois: the continuous battle whereby units are constantly fed into a battle, maintaining its momentum until it finally breaks through the enemy defences (a stance held by d'Urbal); a slow, methodical battle whereby troops advance on small, realizable objectives in a series of limited offensives (Foch's preferred methodology and an important precursor to operational art); and finally the notion that in spring 1915 the French simply did not have the materiel to wage effective operations and should therefore wait for more weapons, especially heavy artillery, and greater expertise (an opinion held by the more conservative commanders Pétain and Fayolle). Seen independent of other factors, d'Urbal's desire to continue attacking 'with the greatest speed, because the more we wait the more the enemy will reorganize themselves and reinforce themselves with men and cannons' seems to make perfect sense.[67] If the enemy is reeling, one should continue to strike. However, even a cursory examination of the operational realities of Second Artois would expose the weakness inherent in d'Urbal's methodology.

The success of May 9, a thoroughly incomplete success at that, was only possible after six days of continuous and well-targeted artillery preparation. The Germans, on the other hand, needed almost no preparation to occupy their rear trenches and fire indiscriminately into the space through which advancing French forces would have to cross. Such fire inflicted gross casualties on the dense masses of advancing French reserves, which one would expect to come across in a thick, continuous stream in a 'continuous' battle. The ease with which the Germans could inflict casualties on advancing French reserves was the principle reason the

64 Palat, *La grande guerre sur le front occidental*, p. 158.

65 *AFGG*, *Tome III*, p. 42.

66 Ibid., p. 47: «*toutes les attaques vont* échouer*; nos troupes sont fatiguées, les unités mélangées; de plus, l'ennemi a fait venir de Lens et de Douai de nouveaux renforts*».

67 Ibid., pp. 47–48: «la plus *grande rapidité, parce que plus nous tarderons et plus l'ennemi se retranchera et amènera contre nous des hommes et des canons*».

Germans would ultimately contain the French advance. D'Urbal's preference of a continuous battle that would break through the enemy lines was a preference for pure fantasy over harsh, unglamorous reality.

A more sensible operational style was espoused by Foch, who opted for a slower, more methodical approach to combat. Concerning the fighting on the crest of Notre Dame de Lorette in the months before Second Artois, Foch said 'I have insisted that it is less important to advance rapidly, *but essential to move securely*, step by step, as each objective is gained.'[68] Foch tended to favour an 'operational' approach throughout Artois, although it is unclear whether or not he still expected a breakthrough at some point. Whereas on the one hand Foch can be quoted demanding a slow and methodical approach, there are instances when he at least references the breakthrough as a possibility: 'breakthrough remains doubtful as long as we have not allocated to the offensive large quantities of the new means [heavy artillery, light machine guns etc.]'.[69] Foch can also be found saying: '[as] masters of *cote* 140 [Vimy Ridge] we could have undertaken [further operations] under good conditions. But the breakthrough was from then on [after failing to secure Vimy] unrealizable', which seems to imply that Foch believed the breakthrough to have been a possibility during Artois.[70]

Foch comes out with a fairly positive reputation during this period. Marshall-Cornwall's *Foch as Military Commander* is on the whole very supportive of his doctrinal and operational management as leader of the GPN; most other scholarly works agree. This is in contrast to how many of Tenth Army's corps and divisional commanders felt: they were often sharply critical of Foch, accusing him and d'Urbal together of setting forth unrealizable goals given the means at Tenth Army's disposal.[71] That said, such accusations were thrown at Foch typically for no other reason than the fact that he was commander of the *Groupe Provisoire du Nord*. He represented a far-off authority at which mid-level commanders could vainly shake their fist. Such complaints never seemed to consider the operational freedom he allowed d'Urbal, nor the necessary role of higher level officers as 'thrusters' to keep operations from bogging down prematurely. During this period Fayolle was regularly frustrated with what he considered to be Foch and d'Urbal's wishful thinking in terms of operational goals, and it is likely that Pétain felt the same. Whether or not Second Artois would have happened at all if they were in charge is debatable.

After nearly a week of fighting it became obvious to all those involved that the offensive had stalled. On 15 May Pétain had wanted to make a push on to Souchez regardless, calling into question his famously conservative style, but was overruled by Foch, who decided that success:

---

68 Marshall-Cornwall, *Foch as Military Commander*, p. 150.

69 *AFGG, Tome III*, p. 108: *«la percée reste douteuse tant que nous n'aurons pas doté l'offensive de beaucoup de moyens nouveaux»*.

70 Ibid. «Maîtres de la hauteur 140, nous *pouvions l'entreprendre dans les bonnes conditions. Mais la percée* était *désormais irréalisable*.»

71 Palat, *La grande guerre sur le front occidental*, p. 341.

> could not be made by a rushed offensive, but by a complete and detailed preparation analogous to that which preceded the attack on 9 May. This preparation requires *time* … furthermore it is necessary to constitute ourselves on a *base de départ* … by actions in front and in flank, at suitable distances, by heavy artillery and by field or trench artillery. All precautions will be taken to render this preparation effective.[72]

Thus Second Artois entered its second phase of small, localized offensives to gain key points of terrain in preparation for a second general offensive on the same scale as 9 May. Doughty, in particular, considers Foch's intervention and call for more detailed and better prepared operations to be commendable.[73] While Foch's call for more thorough preparation was well meaning, the execution was still left to d'Urbal and his corps commanders. As a result, things progressed in much the same manner as before: attacks were poorly supported and lacking in lateral liaison, thereby allowing the Germans to pour flanking fire upon localized attacks with near impunity.

These small, local attacks continued through late May and into June, with the French capturing several key points (*la sucrerie* and the chapel of Notre Dame de Lorette, among others). These attacks, while successfully capturing their objectives, came at incredibly heavy cost for the French, who were at the mercy of ceaseless German artillery bombardments. French troops endured artillery fire during the march to the front and the preparation for the attack, only to be met with yet heavier counter-barrages and machine-gun fire once the infantry went over the top. It is perhaps the only point of absolute concurrence to be found among all of the historians of this battle: the period of small, local attacks was too costly in casualties, which far outweighed the minor tactical advantages earned by capture of their *base de départ*. The French Official History depressingly states that 'in all, despite the valour of our troops and after a month of fierce combat, we were not successful'.[74] What has yet to be addressed is whether or not the local attacks were bound to fail or if, with a bit more skill and patience, these partial offensives could have been less costly and more successful. This question will be addressed when this book examines more closely the maelstrom that lasted from 12 May to 15 June.

By mid-June Tenth Army had captured most of its *base de départ* objectives and was poised to launch its second general offensive. Reinforcements were sent

---

[72] *AFGG, Tome III*, p. 58–59: «ne *peut plus* être *le fait d'une offensive brusquée, mais bien d'une préparation complète et minutieuse, analogue à celle qui à précédé l'attaque du 9 mai. Cette préparation demande du* temps … *D'autre part, il est nécessaire de se constituer une* base de départ … *par des actions de front et de flanc, à bonnes distances, d'artillerie lourde et de campagne ou de tranchée. Toutes précautions seront prises pour rendre cette préparation efficace.*»

[73] Doughty, *Pyrrhic Victory*, p. 162.

[74] *AFGG, Tome III*, p. 71: «en résumé, *malgré la valeur de nos troupes et après un mois de combat acharnées, nous n'avons pas réussi*».

to Tenth Army in the form of infantry divisions and artillery batteries, including several important heavy artillery batteries. While the secondary sources make it appear that this was a large effort to give Tenth Army numerical superiority over the Germans it, in effect, only replaced the battered divisions that had already seen action so that the number of infantry and artillery did not see significant change. Whereas on 9 May the French had 1,075 guns (782 field artillery and 293 heavies), they had 1,160 guns (805 field artillery and 355 heavies) on 16 June.[75] The greatest material difference was seen in the number of shells per gun. On 9 May each 75mm gun was allotted 400 rounds (although Pétain limited his to 300 to build up a reserve); on 16 June d'Urbal provided 800 shells for each 75 mm.[76] In all the French fired 265,430 shells from 3 to 9 May; from 10 to 16 June the French fired 497,122.[77] Yet, when the French attacked from 16 to 18 June they met with complete failure time and time again. What happened?

The French themselves believed that their preparation had not been sufficient, that they had not enough materiel to break the German defences which had been built up over the preceding six weeks since 9 May. They also blamed the failure on an increase in the number of German heavy artillery pieces *vis-à-vis* their own.[78] Neither Doughty nor Clayton ventures any explanations for the renewed French offensive performing as poorly as it did. Palat, alone, specifically blames poor methodology in conjunction with insufficient materiel as a cause for the underperformance of the French in Artois.[79] By June 1915 the paradigm of combat on the Western Front had already changed irrevocably. The Germans had strengthened their defensive positions by digging more complex positions in increasing depth and organizing pre-planned artillery barrages; the Germans were preparing for the long haul. With this development the operational method that had so nearly yielded a major success to the French army had become obsolete. It would require serious revision before the French army could hope to have another chance of achieving lasting results on the battlefield.

With so little written about Second Artois there are left a great many questions to be explored within the body of this work. What were the causes, implications and influences of the feats achieved by Pétain's XXXIII CA on 9 May? How did the influence of higher ranking generals (Foch and Joffre) affect the way the battle was fought? To what extent were the French under-armed; or, conversely, were their methods simply not up to the task?

Fortunately, the paucity of secondary source material on Second Artois is compensated for by a wide range of primary source material. There are a great number of documents in the *Service historique de l'armée de terre* archive at the Château de Vincennes which relate directly to Second Artois – perhaps too

---

75 Ibid., p. 99 and Doughty, *Pyrrhic Victory*, p. 163.

76 *AFGG, Tome III*, p. 89.

77 Ibid., p. 100.

78 Ibid., p. 107.

79 Palat, *La grande guerre sur le front occidental*, pp. 332–337.

many. There are several cartons full of documents dealing with operations and intelligence for each of the seven CAs which took part in Second Artois, and similar information for the 14 infantry divisions that did the bulk of the fighting. The cartons pertaining to XXXIII CA and its constituents, the 70^e^, 77^e^ and Moroccan divisions, are all very detailed and meticulously kept. These were absolutely vital for addressing the central event of the offensive: XXXIII CA's advance on 9 May. Similarly, XXI and XX CAs both have very detailed records concerning this period. XXI CA maintains its stellar depth of information, even going back into earlier fighting on Notre Dame de Lorette (First Artois).

Unfortunately, not every CA maintained such thorough records. The peripheral IX and XVII CAs both suffer from a serious lack of relevant documentation in their cartons. The document catalogue at Vincennes suggests that since XVII CA's carton is essentially barren one might try the relevant brigade-level cartons. According to this same volume there are no brigade-level cartons available for XVII CA. My understanding of the actions of IX CA have been based almost entirely on the documents stored in the carton of its 58^e^ and 17^e^ DIs, both of which were so rich that they hold documentation relevant to the entire CA, and give a remarkably clear picture of operations on the French left, which was adjacent to the British army. The situation of XVII CA's carton is similar, with the divisional records making up for the lack of corps-level records.

Ultimately, it is the objective of this work to explain the full importance of the French effort in Artois from May to June 1915: a period which saw the first widespread and doctrinally supported use of infiltration tactics and the rolling barrage; the first Allied use of poison gas; a rapid and major advance; and an engagement upon which trench-warfare doctrine for both the French and Germans would rest for much of the war. A range of key issues will be addressed in an attempt to ascertain how such issues might have affected the battle's outcome and, furthermore, how French commanders came to view the battle and its perceived lessons. At the same time this book will explore spring 1915 more generally as a period of rapid evolution and intense interest to the study of the First World War. Thus, in addition to detailing the events of Second Artois this book explores the role of GQG, especially as it pertains to the dissemination of doctrine and the management of the French army as a whole. The artillery will be covered in a chapter on its own, not only in recognition of its primary importance on the First World War battlefield but also in order to better illustrate the material evolution of the French army up to and through spring 1915. Finally, a case study of two divisions (77^e^ and 34^e^) will perform a similar function for the infantry, allowing for a more detailed look at infantry practices, variability and evolution in this crucial time period. It is hoped that, using Second Artois as a case study, the whole of 1915 might be brought more towards the forefront of historical consciousness, to encourage others to seek to better understand this year and its larger significance to the evolution of the Great War.

Chapter 1
GQG and the Changing Doctrine of Trench Warfare

In order to place Second Artois in its full context, and to better illustrate the intellectual and evolutionary state of the French army in 1915, it is necessary to discuss the role played by *Grand Quartier Général* (GQG). GQG, the headquarters of commander-in-chief Joseph Joffre, was the highest administrative body in the French army. It was at GQG that ideas coming up from the front were digested, refined and then circulated through the entire army. This process of reflection would produce the new and innovative doctrine on which Second Artois would be based, and later process the lessons learned during the battle for use in future engagements. GQG would not only help refine and direct the flow of new ideas circulating around the army, but would also help orchestrate the production of the new equipment which trench warfare demanded: from guns and ammunition to telephone wire and gas masks. The use of new materiel was in many ways as important as the contemporaneous doctrinal evolution taking place in 1915, and helps underscore the magnitude of the change that 1915 brought to the Western Front. GQG would also act as an important liaison with the British army as the Allies fought their first combined trench battles – the first tentative steps of a partnership which would ultimately win the war.

Spring 1915 saw GQG attempt to achieve an unprecedented level of influence over the tactics and operations conducted by its executors in the field. Facilitated by the trench stalemate, which both eased communication difficulties and posed a serious problem that demanded solving, the men at *Grand Quartier Général* would begin a process of vertical integration in the hope of creating a single tactical system that could break the trench deadlock and drive the Germans from French soil. This system, while it would not reach full maturity in 1915, would be the basis for the French successes in 1916 and afterwards. The British were not free from this trend towards uniformity and centralization. The French, who at this time were materially, tactically and politically far stronger than their British allies, would do their best to conduct strategy as if the British were not a separate, independent army but merely a component part of the French war effort.

The high command of a modern fighting force has a variety of roles and functions far too myriad to be covered in the course of a single chapter. High command must interact with government officials for a number of reasons, either as a unit or individually. Such interaction could be for the coordination and approval of grand strategy; or it might be in an attempt to secure greater support and funding for certain military projects, or any number of other situations which demand the cooperation of civilian and military power structures. High command is usually one of the principal originators of propaganda, which might be directed at the troops, civilians or both. This, yet again, often provides an opportunity for members of the high command and politicians to interact. High command must liaise with political bodies (in France during the Great War it was the War Ministry) to deal with private industrialists who build the many tools, lethal and otherwise, that modern armies require. Furthermore, members of the high command will typically be in a position to approve or disapprove propositions for new designs and new weapons. High command is responsible for managing the material and human resources allotted to it, a function which includes everything from the provision of foodstuffs to the training of men and officers; it also provides the means, the mentality and the motivation for organized conflict in the field. In other words, high command provides an army with military hardware, doctrine, resource management, propaganda and strategy; it is the brain of an army.

As adequate coverage of all of these vital functions is not feasible within the context of this work, this chapter will focus primarily on GQG's role as the disseminator of official doctrine and interlocutor with the British army. Both offer considerable insight into the dynamic French conduct of war as it adjusted to the new, gruelling conditions of trench warfare without resorting to a basic analysis of guns and shells. We will begin by looking at the doctrinal *redressement* as documented and disseminated by GQG.

Writing a Doctrine for Trench Warfare

Spring 1915 was one of the most important periods of activity in the forming of doctrine, thus addressing one of the weak points of the French army before 1914. Joffre recognized this failing of the French in 1911 when he became the *chef d'état-major général*, stating in his memoirs that 'First off, it was necessary to give our army a firm doctrine for war, known by all and unanimously accepted.'[1] His success in doing so is the subject of intense debate. Douglas Porch asserts that the 'post-1900 French army was simply not capable of formulating or applying a tactical doctrine'.[2] Instead, he claims that the maligned *offensive à outrance*[3] was not a doctrine at all, but merely a weak attempt to motivate soldiers trained in a system which lacked any solid tactical precepts. Robert Doughty is a bit more

[1] Joffre, Maréchal (1932), *Mémoires du Maréchal Joffre (1910–1917)* (Paris: Librarie Plon), p. 29: «Avant tout, *il fallait doter notre armée d'une doctrine de guerre ferme, connue de tous, et unanimement acceptée.*»

[2] Porch, Douglas (1981), *The March to the Marne: The French Army 1871–1914* (Cambridge: Cambridge University Press), pp. 214–216.

[3] The offensive 'to the limit' or 'to the extreme'. The *offensive à outrance* was a guiding principle of the French army as it marched to war in 1914, which encouraged French commanders to be on the offensive at all times, regardless of the situation or the costs.

generous in at least considering the *offensive à outrance* to be an actual doctrine.[4] Michel Goya cites several doctrinal documents, yet must ultimately reject the idea that such a body of work really represented an integrated and functional doctrine 'unanimously accepted'.[5] Of course, had GQG been more successful in creating and disseminating a coherent doctrine the advent of trench warfare on the Western Front would have rendered it obsolete anyway. One might also posit that had a highly codified official doctrine been in place it could have made a quick adaptation to trench warfare more difficult. As a comparison, the generalized nature of Plan XVII[6] had built-in flexibility which allowed the French to effectively shift the weight of their forces in the first manic weeks of the war. Had the French failed to do so, they would not have been in a position to counter the initial German successes, and may well have lost the war.

Already in spring 1915 GQG could quickly and reliably receive transmissions from the front, digest them and mould doctrine accordingly. This was achieved despite the lack of standardized forms and practices which now seem organic to large bureaucracies. The vast majority of the after-action reports produced in 1915 were still being scribbled on scraps of paper in the front-line trenches and passed up to regiment or brigade headquarters via runners. From there the reports would normally be telegraphed or typed and delivered to the divisional and corps commanders who would then collate the reports and send them on to the relevant army commander. Eventually, reports which had reached the commander of an army or army group might be communicated directly to GQG via telegraph, telephone or delivered in person. During periods of high stress, such as the weeks comprising Second Ypres and leading up to the launch of Second Artois, Foch spoke over the phone with GQG every single day, often several times a day.[7] By mid-1915 an increasing number of *compte-rendus* (reports) coming from front-line units were either hand-written or typed on standardized forms, or were at least given a uniform structure. While this system was not perfect (GQG had more than once to ask that reports be more detailed), it still facilitated a more efficient flow of information up the chain of command.[8]

The first major result of GQG's receipt of an ever-increasing amount of data from lower level officers was *But et conditions d'une action offensive d'ensemble* (Goal and Conditions for a General Offensive Action, or Note 5779). This was a monumentally important work of doctrine designed as a single reference point

4 Doughty, *Pyrrhic Victory*, p. 26.

5 Goya, *La chair et l'acier*, pp. 110–112.

6 Plan XVII was the French troop concentration plan in the event of war with Germany, finalized in May 1914. As it lacked pre-set strategic plans it allowed Joffre to respond to situations as they occurred at the outbreak of war, a flexibility which helped thwart the German drive into France later that year. Doughty, *Pyrrhic Victory*, p. 37.

7 See *SHD*, 16N1807.

8 *SHD*, 16N1677; «Message *Téléphoné pour Monsieur le Général Foch pour Monsieur le Général Commandant l'Armée à Saint-Pol» 3 juin 1915; no. 3096.*

for field commanders engaged in offensive operations. The document was not so much a piece of original doctrine as it was an amalgamation of the individual lessons learned piece-meal in 1914 and early 1915. These lessons were processed and bound together to form a singular work which would be the bedrock of an accessible and coherent doctrine for the French army. The hope was to spread quickly the methodology of more effective units across the entire army, thus making French operations more effective and efficient along the entirety of the Western Front.

Of course, how widespread the document's influence eventually was is debatable. Doctrine can sometimes be hard to trace down to specific actions at the front. Furthermore, it is not always possible to differentiate between official doctrine and widespread 'best practices' which arose of their own accord. Nevertheless, Note 5779 is a highly compelling document for historians of the Great War as it captures the essence of 1915 in which pre-war idealism (such as the *offensive à outrance*), pushed by 'Young Turks' at GQG, coexisted with a realism earned in combat and based on a reliance on artillery preparation and clearly defined, easily attainable objectives.

The very first sentence of the document reads as follows: 'The goal of an offensive action is not only to take a line of enemy trenches, but to also chase the enemy *as a whole* from his position and *defeat* him without giving him time to re-establish himself.'[9] It is made perfectly clear that each offensive should have as its goal a breakthrough and rout of the enemy. This idealism is immediately followed by more grounded advice: 'an … operation does not improvise itself. Repeated experiences have shown that a detailed organization and preparation are the indispensable conditions of success. This preparation ought to leave nothing to chance and realize, before the launching of an attack, all the conditions favourable to its execution.'[10] The need for a methodical preparation was being highlighted by many influential figures in the French army in 1915, including Foch, Pétain and Joffre. A letter from GQG's 3e Bureau[11] to French Army commanders, composed six days before the release of *But et conditions d'une action offensive d'ensemble*, stated:

---

[9] *SHD*, 19N735; «But et conditions et *d'une action offensive d'ensemble*»: «Le but *d'une action offensive n'est pas uniquement de s'emparer de une ligne de tranchées adverses, mais bien de chasser l'ennemi* de l'ensemble *de sa position et* de le battre, *sans lui laisser le temps de se rétablir.*»

[10] Ibid. «*une … opération ne s'improvise pas. Des expériences répétées ont montré qu'une organisation et une préparation minutieuses sont les conditions indispensable du succès. Cette preparation doit ne rien laisser au hasard et réaliser, avant le déclanchement de l'attaque, toutes les conditions favorables à son exécution.*»

[11] 3e Bureau was the 'Operations' bureau and was in charge of doctrinal information, among other things. The accusation of 'Young Turks' in the French army were generally aimed at 3e Bureau.

> In our recent operations our infantrymen have everywhere attacked with a remarkable spirit, but have only succeeded in those actions [which have been] rigorously and methodically organized: where the infantry are, beforehand, established at a good attacking distance, [when attacks] have not been launched except after a complete destruction of the enemy's accessory defences [i.e. barbed wire] and a bombardment of their first and second line trenches; where liaison of every sort is assured; where the attack, developed along a sufficiently [long] front could be sustained and prolonged by the arrival of reinforcements, [and] supported by artillery throughout its duration.[12]

Comparable to the above is Foch's mantra of 'slow, methodical and prolonged' preparation, or Pétain's detail-oriented leadership on the battlefield, which shall be covered later.

After elaborating briefly on the tasks and responsibilities of commanders and their units when preparing for an attack, the note declares two operational necessities whose coexistence would be problematic, to say the least. At the top of page two the note states that 'an attack ought to be launched along an extended front to obtain significant results and also to stop enemy flanking attacks'.[13] Experience had shown that many of the small-scale attacks launched in First Champagne (February–March 1915) suffered horrific enfilade fire. This GQG hoped to minimize by extending the front as much as possible.[14] Attacking along an extended front would have the added benefit of preventing the Germans from concentrating their reserves on only one small portion of the front, thus increasing the chances of a tactical breakthrough. General Fernand de Langle de Cary, commander of Fourth Army during the recent battle in Champagne, had felt that success across the whole of a front could not be achieved. However, partial successes were to be expected and if these partial successes were exploited laterally they could create a hole large enough through which a strong exploitative attack could be launched.[15] This was yet one more reason to attack on a broad front as it would, in essence, present more opportunities for achieving a local breakthrough.

---

12 *SHD*, 18N94; «Note pour les Commandants des *Armées (Personnelle et Secret)» 10 avril 1915*: «*Dans nos récentes opérations, nos fantassins ont partout attaqué avec un entrain remarquable. Mais, seules, ont réussi entièrement les action rigoureusement et méthodiquement organisées: où l'infanterie, préalablement* établie *à bonne distance d'assaut, n'a* été *déclenchée qu'après destruction complète des défenses accessoires et bouleversement des tranchées ennemies de première et seconde lignes; où les liaisons de toute nature* étaient *assurées; où l'attaque, développée sur un front suffisant, a pu* être *soutenue et prolongée par l'arrivée des renforts, appuyé par l'artillerie pendant toute son développement.»*

13 *SHD*, 19N735; «But et conditions *d'une action offensive d'ensemble*» 16 *avril 1915*: «*l'attaque doit* être *mené sur un front* étendue *pour obtenir des résultats important et supprimer les flanquements ennemis*».

14 Doughty, *Pyrrhic Victory*, pp. 133–134.

15 *AFGG, Tome II*, p. 414.

The very next sentence in Note 5779 states that 'the large units initially engaged (army corps, divisions) will be side-by-side; each of them operating on a narrow front and arrayed in depth, to ensure the *continuity* and strength of the action.'[16] This statement is to be considered as the dispositional prerequisite for the pursuance of the note's stated goal for offensives: the *percée*. The relationship is unambiguously stated not much later in the document with the phrase: 'The executors, at all levels, will be infused with the idea of the breakthrough … to continue the attack without stopping until the final result.'[17] The idea of a 'continuous battle', as it was called, was certainly not new.[18] Even before First Champagne it was considered necessary for the assault troops to be followed very closely by their reserves, who could further exploit the success of the troops in front of them.[19] This hypothesis was apparently vindicated in February 1915 when reserve units 3 or 4 kilometres behind the front line could not provide any real support to the foremost attacking units. When the above two statements are taken together in the same document, however, they pose a near impossible quandary. Tactical commanders were expected by their superiors to maintain a strong attacking posture both laterally and in depth while on meagre rations of men and materiel. This clear contrast in operational methodology sharply underscores the inexperience and indecisiveness of the officers in 3e Bureau. Their recipe for success in 1915 was one based on size and mass: more guns and more men. With such a methodology every attack was going to be costly.

The note departs from this simplistic formula and goes on to provide helpful advice pertaining to the need for close artillery–infantry liaison and for the importance of sheltering infantry who are waiting to attack, so as to minimize unnecessary casualties. The note also reiterates the importance of a complete preparation, and the duty of officers to ensure that the enemy position has been sufficiently bombarded to warrant an attack.[20] The main body of the document ends with a reaffirmation of the continuous attack supported by artillery throughout its duration and states rather vaguely that only an 'energetic, resolute' commander is fit to command offensive operations.[21] Thus, in only four pages the tactical art of the French army in 1915 was summarized and codified. Attached to this note were several annexes which provide a more in-depth look at various topics of interest concerning the actual execution of an organized trench battle. These annexes

---

16 *SHD*, 19N735; «But et conditions *d'une action offensive d'ensemble*» 16 *avril 1915*: «Les *grandes unités initialement engagées (Corps d'Armée, Divisions) seront accolées; chacune d'elles opérant sur un front* étroit *et articulée en profondeur, pour assurer la* continuité *et la puissance de l'action.*»

17 Ibid. «Les *exécutants, à tout les* échelons, *seront pénétrés de l'idée de percer … de continuer l'attaque sans arrêt jusqu'à résultat final.*»

18 Doughty, *Pyrrhic Victory*, p. 131.

19 Palat, *La grande guerre sur le front occidental*, p. 92.

20 *SHD*, 19N735; «But et conditions *d'une action offensive d'ensemble*» 16 *avril 1915*.

21 Ibid.

would have a demonstrable effect upon the conduct of offensive action in 1915 owing to their clarity and practicality, most especially those sections which deal with artillery and infantry tactics.

The Artillery Annex

The first mention of the role of artillery in Note 5779 states that:

> The artillery will, by a *systematic fire on the successive lines of the enemy*, destroy the accessory defences, the dug-outs, their communication trenches and their flanking positions; it will demoralize the enemy by repeated shocks and will be prepared to support infantry attacks effectively for their duration, not just on their first objective but on their further objectives [as well].[22]

This statement is a clear definition of the duties of artillery which reflects how it was actually being used on the battlefield. This is somewhat contrasted by the first sentence in the artillery annex, which reads: 'The will to always push forward by continued action compels the artillery to have a very complete *initial organization* and to *minutely study subsequent displacements*.'[23] It is worth noting that 'willpower', which itself is only a byword for the 'offensive spirit', comes to the forefront of what is expected from the artillery. This is especially interesting in light of the fact that the 'cult of the offensive' tends to be applied to the infantry and cavalry rather than technical arms such as the artillery.[24] However, pre-war doctrine, which expected the artillery's main role to be direct infantry support, would naturally lead the artillery to take up forward positions from which they could easily observe the infantry action and intervene accordingly.[25] This is largely another example of the integration of pre-war notions on the conduct of war with assessments founded on experience of the actual battlefields on which artillery was

---

22 Ibid. «*L'artillerie aura dû, par* <u>*un tir systématique, sur les lignes successives de l'ennemi,*</u> *détruire les défenses accessoires, bouleverser les abris, ses boyaux des communication et ses organes de flanquement; elle aura déprimé par des chocs répétés le moral des défenseurs et sera préparée à appuyer efficacement les attaques de l'infanterie pendant tout leur durée, non seulement sur son premier objectif, mais sur les objectifs ultérieurs.*»

23 Ibid. «La *volonté de pousser toujours plus avant par une action continue, entraîne pour l'artillerie une* <u>*organisation initiale*</u> *très complète et l'*<u>étude *minutieuse de ses déplacements*</u> *ultérieurs.*»

24 Goya, *La chair et l'acier*, pp. 56–63 and House, Jonathan M. (1984), 'Toward Combined Arms Warfare: A Survey of 20th-Century Tactics, Doctrine and Organization', *Combat Studies Institute*, Research Survey No. 2, pp. 164–169.

25 Ripperger, Robert M. (1995), 'The Development of French Artillery for the Offensive, 1890–1914', *Journal of Military History*, vol. 59, no. 4, p. 601.

being deployed. While there were good reasons to station some artillery batteries close to the front lines (if nothing else it facilitated fire deeper into the enemy rear defences), it was not normally done with the immediate intention of readying the artillery to follow in the wake of a successful infantry attack. In the artillery annex the old notion of the dash of the field (or horse) artilleryman constantly displacing his battery to support forward infantry is thus maintained. This was an exercise in undue optimism owing to the rarity of major forward displacements undertaken by the artillery in 1915.

Another interesting assertion made by the annex refers to the appropriate use of the quick-firing French 75mm light artillery piece, the most common gun available to the French army. Note 5779 recommends that: 'The speed of fire of the 75mm ought never to exceed four rounds per minute and fire of this speed should never last for more than 10 minutes at a time.'[26] This recommendation is baffling when one considers the fact that the 75 could fire up to 15 rounds a minute, and was designed to blanket large areas with shrapnel over a short period of time. Asking that the French 75 be fired at only four rounds per minute would place it at the same cadence as the outmoded de Bange series of guns from the 1870s–1880s, and considerably slow down artillery preparation.[27]

The reason for this drastic recommendation was concern regarding the over-use and subsequent degradation of the French army's signature gun. Simple wear and tear had gradually reduced the number of 75s from 4,170 in August 1914 to 3,365 by the time Second Artois began.[28] Production quality and quantity dropped substantially when the French lost its industrial north-east to the Germans in 1914, making it increasingly difficult for GQG to meet the demand of the French army for guns and munitions as the *materialschlact*[29] intensified.[30] As the supply of new guns coming off the production line dipped in 1914–15 the guns lost in the field, owing to a mixture of simple wear and tear and sheer negligence, could not be readily replaced.[31] This posed a major threat to the French ability to wage war. Indeed, losses owing to barrel-wear and other mechanical failures would far outstrip any losses suffered as a result of enemy fire throughout Second Artois. Regardless of the potential danger of overworking their guns, commanders in the field (most notably General Maistre, commander of the XXI *Corps d'Armée*,

---

26 *SHD*, 19N735; «But et conditions *d'une action offensive d'ensemble*» 16 *avril 1915*: «La *vitesse du tir de 75 ne devra jamais dépasser 4 coups à la minute et le tir avec cette vitesse ne devra pas* être *prolongé sans arrêt pendant plus de dix minutes.*»

27 Goya, *La chair et l'acier*, pp. 148–155.

28 Daille, *Histoire de la Guerre Mondiale*, p. 110.

29 German for 'material battle', the way in which Germans saw the materiel intense fighting in the First World War.

30 For more see Porte, Rémy (2005), *La mobilisation industrielle: «premier front» de la grande guerre?* (Cahors: 14–18 Éditions).

31 Campana, J. (1923), *Les progrès de l'artillerie: l'artillerie française pendant la guerre 1914–1918* (Paris: Imprimerie la Renaissance), p. 43.

CA, which fought over Notre Dame de Lorette) openly disregarded the constraint of firing only four rounds per minute and chose to continue to use their 75s as deemed necessary.[32]

The annex assigned the artillery four main tasks: the destruction of enemy defences, anti-personnel fire, counter-battery fire and direct infantry support during an attack.[33] The expansion of the role of artillery from a merely supportive arm to a preparative arm was crucial to success in the Great War. The list above also compares favourably with the list of five tasks for artillery as laid out by Andre Laffargue in his Étude *sur l'attaque dans le période actuelle de la guerre*, published in mid–late 1915: Laffargue made a separate point on the importance of taking out enemy machine guns, which Note 5779 simply lumped in with 'the destruction of enemy defences'.[34] The annex also placed special emphasis on the importance of aerial reconnaissance and assigns appropriate targets and bombardment durations for heavy artillery, field artillery and mortars.

Of particular interest is section B of the annex, which deals with artillery action during an attack, as opposed to in preparation. The document states that 'at the hour fixed for the infantry attack the artillery will increase its range *progressively* to make, in front and on the flanks of the attack, a longitudinal and transversal barrage to shelter the infantry so they can advance'.[35] This statement represents one of the earliest and most succinct descriptions of a rolling barrage in the First World War, and is especially important as it comes in the form of official doctrine as early as the spring of 1915. Nevertheless, there remains considerable debate about just when the first rolling barrage occurred. Michel Goya gives credit to the 33e *Division d'Infanterie* (DI), whom he claims used a rolling barrage in August 1914; Robert Doughty gives a more conservative answer, saying that it was first used by Fourth Army in October of 1915; others still attribute its first manifestation to General Nivelle during Verdun.[36]

In some ways this argument hinges upon one's definition of a rolling barrage. All can agree that a rolling barrage consists of a bombardment which, on a set timetable, increases the range of its fire to act as a shield for the advancing infantry as they move through the enemy's defences (thus differentiating it from a traditional barrage, which would only be fired on a static target area). This moving wall of

---

32 *SHD*, 22N1472; «Le Général Maistre Commandant le 21e *Corps d'Armée à M.M. les Généraux Cdt les, 13e, 43e, 48e, 58e, et 92e Divisions, le Général Cdt l'Artillerie*» 22 *juin 1915*.

33 *SHD*, 19N735; «But et conditions *d'une action offensive d'ensemble*» 16 *avril 1915*.

34 Laffargue, André (1916), Étude *sur l'attaque dans la période actuelle de la guerre: impressions et réflexions d'un commandant de compagnie* (Paris: Plon-Nourrit et Cie), p. 8.

35 *SHD*, 19N735; «But et conditions *d'une action offensive d'ensemble*» *16 avril 1915*: «*à l'heure fixée pour l'attaque de l'infanterie, l'artillerie allonge progressivement son tir, pour faire, en avant et sur les flancs de l'attaque, un barrage longitudinal et transversal à l'abri duquel l'infanterie peut progresser*».

36 Doughty, *Pyrrhic Victory*, p. 194 and Goya, *La chair et l'acier*, p. 190.

fire and steel, if executed properly and closely followed by the attacking infantry, would force enemy defenders to remain in cover (usually in the form of deep dug-outs) right up to the moment when the attacking infantry were upon them, thus leaving them little or no time to emerge out of their bunkers and provide an organized resistance. What Goya claims was the first rolling barrage was likely an execution of *tir progressif*, a manoeuvre in which the artillery would fire four sets of shells (normally not more than two or three per set) at increasing ranges in quick succession to create a sort of sweeping effect. This method was in common use across the French army until January 1915, when Joffre ordered that the practice be stopped as it was too costly in terms of shells. It would, nevertheless, remain standard practice for anti-aircraft fire.[37] In any case, there does not appear to be doctrinal support for a rolling barrage in any documentation before Note 5779, which makes its particular iteration of this key tactic particularly noteworthy.

According to the annex, the artillery was meant to protect the infantry after it had captured the first enemy trenches by direct fire support. This would allow the infantry to repair the captured German trenches, making them ready to defend against any counter-attacks. A pause in the advance would also give the artillery a chance to further its preparation of the second and third German positions, paying special attention to the destruction of machine guns which had revealed themselves during the course of the attack.[38] The advent of a slower, more methodical approach to offensive action, as opposed to the continuous offensive aspired to earlier in the document, is significant for demonstrating the competition between the two notions of how a battle should be fought. Once the infantry began its attack a certain number of batteries were also expected to advance along with the infantry, occupying pre-prepared sites in order to maintain direct support for the most forward advancing units.[39] This tactic would be highly successful for the French during the Battle of the Somme in 1916.

The final sections of the artillery annex cover counter-battery fire and infantry–artillery liaison. The annex states that the 75mm should be the preferred gun for counter-battery fire, but if the enemy guns are located out of range they could be attacked with heavy artillery. Counter-battery fire is called 'one of the primary obligations' of artillery and should aim to silence as completely as possible all enemy batteries which will be located via terrestrial and aerial reconnaissance.[40] On the subject of reconnaissance the annex is adamant in stating that artillery cannot operate effectively without good observation and a close liaison with the infantry. Unlike many other documents the annex actually designates two types of observation post: one for observing and correcting friendly artillery fire, and another for the discovery of hidden enemy targets. The annex goes on to state the

[37] Dubois, Pierre Joseph Louis Alfred (1916), *L'artillerie de campagne dans la guerre actuelle 75 & 90* (Paris: L. Fournier), p. 127.

[38] *SHD*, 19N735; «But et conditions *d'une action offensive d'ensemble»16 avril 1915.*

[39] Ibid.

[40] Ibid.

importance of forward artillery spotters in the accurate observation of fire, thereby breaking with pre-war regulations which relegated all observation to within the immediate vicinity of the guns.[41] Aside from the increased ease in observation afforded by moving artillery spotters closer to their targets, it also facilitated better liaison. The presence of forward artillery observers allowed the infantry to make requests to the artillery directly, in real time, as opposed to having to send a runner who may or may not meet with obliging battery commanders once (and if) he found his way to the rear.[42]

The annex ends with a statement on the use of telephonic communication between the infantry and artillery. The annex claims that such rapid and clear communication as the telephone could provide was crucial to ensure a complete and timely artillery preparation before an attack. As telephonic communication was easily disrupted by enemy fire the annex suggests that all lines be laid in triplicate along the bottom of *boyaux* (communication trenches). Barring the existence of telephonic communication, the annex rests by recommending the usage of 'conventional' methods such as flags and flares which, while both fraught with difficulty and easily impersonated by the enemy with potentially frightful results, would be the only method of liaison available to advancing infantry.

Orchestrating the Break-in

Equally important to the artillery annex is the annex on infantry tactics. This annex immediately claims that an infantry attack should not aim at merely capturing the first line of enemy trenches, but should try to push on to distant objectives beyond the enemy front line: 'the possession of which will affirm that the primary goal – the breakthrough of the enemy line – has been attained'.[43] Thus, again the document oscillated from a 'continuous battle' as recommended in the note proper to a slower, more methodical battle espoused by the artillery annex before finally returning to the idea of the continuous breakthrough battle in the infantry tactics annex. The annex goes on to explain that infantry should progress in dispersed formation acting as skirmishers, moving from cover to cover, while advancing against the enemy, an idea that was largely in line with pre-war thinking.[44]

Perhaps the most interesting part of this section deals with what are called *nettoyeurs de tranchée* or 'trench clearers': specially trained and equipped soldiers who represent the earliest iteration of infiltration tactics in established doctrine.

41 Pre-war regulations in question: FRANCE, *Règlement provisoire de manoeuvre de l'artillerie de campagne* (Paris, 1902), p. 123.

42 *SHD*, 19N735; «But et conditions *d'une action offensive d'ensemble*» 16 *avril 1915*.

43 Ibid. «de *transmettre aux batteries les besoins de l'infanterie*»; «*dont la possession affirmera que le premier résultat à attendre – percer le ligne ennemi – a* été *obtenu*».

44 House, Jonathan M. (1976), 'The Decisive Attack: A New Look at French Infantry Tactics on the Eve of World War One', *Military Affairs*, vol. 40, no. 4, p. 165.

The annex states that the first waves of infantry should avoid enemy strong-points and penetrate as far as they can. Meanwhile, secondary waves made up of *nettoyeurs* would clear the strong-points bypassed by the assault troops.[45] The *nettoyeurs* were specially provisioned with revolvers, trench knives and grenades to equip them for close-range trench fighting, and received special training in this department. They were assisted by engineers who carried their normal tools plus machine guns, petards (explosives), barbed wire, sandbags and trench mortars.[46] Ultimately, infiltration tactics – sometimes called 'Hutier' tactics and typically considered to be of German provenance – would become an integral part of offensive operations in the First World War.[47] It should not be surprising that an army whose mantra was 'attack' would lead the way in offensive innovation, contrary to earlier historiography which upholds the Germans as the sole master of all things tactical. What might be surprising is just how early the concept of infiltration tactics was recognized and made doctrine by the French army.

It appears likely that *But et conditions d'une action offensive d'ensemble* was an influential document, and may even be the single most important document of 1915 in terms of its effect on operations. Unfortunately, it is impossible to ever ascertain fully the precise impact of this document on operational planning. In the run-up to Second Artois General Maistre made several references to Note 5779, but was apparently content to use it only as and when he felt necessary. In one particular instance Maistre defended the preparation of the 17ᵉ DI as being satisfactory, despite pressure from d'Urbal, who felt that the 17ᵉ DI's preparation was insufficient as it did not conform to Note 5779.[48] While Maistre's overt rejection of doctrinal constraints is telling of the inherent problems in trying to impose doctrine on a fighting force it should also be taken into account that the other divisions under his command did in fact adhere to the note verbatim. The divergence of 17ᵉ DI occurred because it had not had sufficient time to construct a trench network which adhered to the new doctrine. 17ᵉ DI had been placed under the command of Maistre's XXI CA only days earlier, causing serious administrative complications and delays in their preparation.

A few kilometres south of the 17ᵉ DI the *Division Marocaine* (DM), which did not arrive at Berthonval Farm in XXXIII CA's sector until 25 April, reported having received copies of Note 5779 in a dossier filled with general information about the front they were about to take over.[49] The men of the DM had already been acquainted with the new doctrine once, while sojourning in Reims during their transfer to Artois. They received further training upon arrival: a response to

45 *SHD*, 19N735; «But et conditions *d'une action offensive d'ensemble*» 16 *avril 1915*.

46 Ibid.

47 Gudmundsson, *Stormtroop Tactics*, p. 173; Millett and Murray, *Military Effectiveness*, p. 20.

48 *SHD*, 22N1472; «Le Général Maistre Commandant le 21ᵉ *Corps d'Armée à M le Général Commandant le 10ᵉ Armée*» 6 *mai 1915*.

49 *SHD*, 24N2908; «Rapport des *Opérations du 9 Mai 1915*».

the importance of their role in the attack (which was the main assault on Vimy Ridge). The extensive training of the DM was further facilitated by Pétain's careful attention to detail and the abundant materiel allotted to XXXIII CA. On the one hand this meant that they had more tools with which to fight, and on the other it meant that they had more complex procedures to master compared to less well-endowed formations.[50] While there are myriad different reasons why a formation might be more or less successful in the field it is still interesting to note that whereas the 17ᵉ DI, which did not particularly follow the new doctrine, failed to produce any appreciable results over the course of the battle, the DM mounted one of the most successful attacks of the entire year.

GQG as Agent of Liaison and Arbiter of Strategy

Just as GQG offered direction to its own troops in terms of their operational art, so too did GQG attempt to direct Allied strategy on the Western Front. French dominance in this realm was asserted very early on in the war. On 18 August 1914 Joffre sent a letter to the British commander-in-chief, Sir John French, stating that he would be grateful if Sir John would confirm that Joffre's strategy met with no major objections in British GHQ.[51] Sir John responded the next day, insisting that the French could count on British cooperation. The French took this to mean more than Sir John French intended. They would consider the British to be not merely a junior partner, but would generally plan strategy as if the British were a component part of the French army. As such, a large amount of the correspondence between Joffre and French was of a mundane and superficial nature, with most of the concrete interaction generally happening between liaison officers working for army or army group commanders. Joffre would either ask or thank Sir John French for his continued cooperation, depending on the circumstance, and Sir John would reply cordially, offering his continued services.[52] This polite formality papered over the many strong personal and professional disagreements the French and British had throughout the war.

When GQG did entertain more serious dialogue with the British in 1915 it was often Lord Kitchener (Great Britain's Secretary of State for War from 1914 to 1916) who the highest French officers dealt with. A series of meetings in late March give a good impression of how the two parties interacted, and how their largely divergent strategic ambitions often gave rise to considerable friction. On 27 March 1915 Joffre expressed his opinions of the current strategic situation in an internal document written just before Kitchener's arrival in France. In what modern minds would see as a clear articulation of 'total war', Joffre stated that in the current war the French and the Germans were both situated, militarily and otherwise, to exert the entirety

---

50 Ibid.

51 *SHD*, 16N1905; «Relations entre les Commandants Français et *Anglais (Extrait de documents)»*.

52 Ibid., various documents.

of their nations' force on each other.[53] As such, Joffre saw victory as coming about only through a long, attritional struggle between these two forces (an idea perhaps unexpected by those who see Joffre as the champion of decisive breakthrough engagements). The document exposes Joffre's contempt of the fact that while Germany and France are engaged in a mortal struggle Britain had thus far fielded an army which was five to six times smaller than either of the other two.[54] In light of the slow build-up of the British Expeditionary Force (BEF) Joffre concluded that the French must organize themselves to out-last their enemies, first and foremost. This was to be done by taking the utmost care in the training and preparation of men; not expecting untrained or partially trained men to hold trenches (thereby exposing them to losses before they could be formed into worthwhile soldiers); and being generally thrifty in the expenditure of men and munitions until the BEF grew large enough to free up sufficient French troops for a large-scale attack (Figure 1.1).[55]

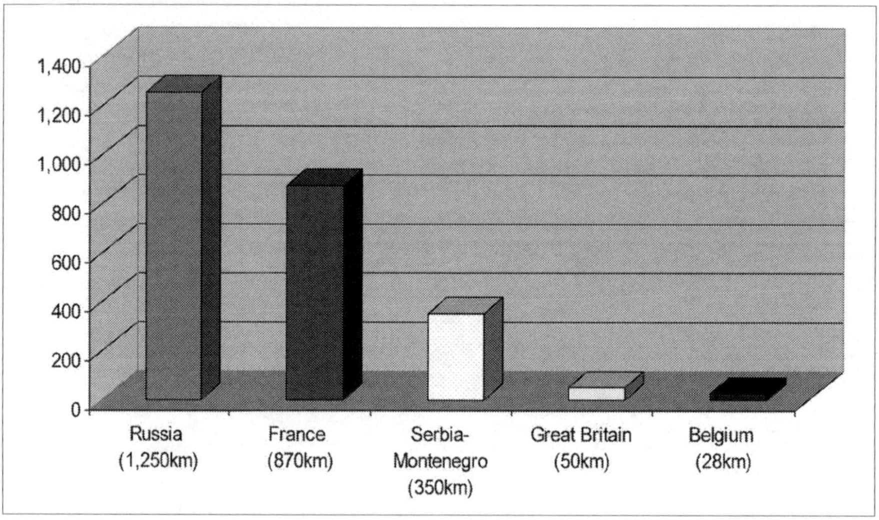

Figure 1.1 Allied Frontages in March 1915
Source: SHD 16N1905, various documents.

During the actual meeting with Kitchener many of the key strategic considerations of 1915 were addressed and plans were laid concerning how best to fight the war. It was decided that while the blockade was certainly helpful it could not possibly win the war on its own, an admission which seemed to imply that if

53 *SHD*, 16N1905; «Note (à propos de la *visite de Lord K.)» 27 mars 1915.*

54 Ibid.

55 Ibid.

Britain was to contribute to the war it needed to be with troops on the ground in France and Belgium.[56] Many of the salient points from Joffre's musings on 27 March came out during the meeting, with the results falling consistently in line with French strategic thinking. Joffre reiterated his prescient belief that the war was 'a siege; thus, long and difficult'.[57] He furthermore made the point that while he considered the French to be just slightly superior to the Germans in terms of men and materiel, a vast superiority was needed to attack successfully in trench warfare.[58] To help the French gain at least a local superiority of some value over the Germans he asked Kitchener to convince Sir John French to relieve the French IX and XX CAs for use in Artois, a request Sir John would ultimately agree to on 1 April 1915. Discussing the need to mass Allied troops against the Germans, Joffre stated unequivocally that the French could only spare one further division for operations in the Dardanelles, and he asserted that any further efforts should be almost entirely naval to spare infantry for the main theatre. Such resistance was to be expected. Joffre had fought vigorously against any campaign in the Near East, and would continue to do so, feeling that any troop diversions would only serve to weaken his position in the west.[59]

Joffre had been given near-limitless power before the war, and this power only grew with the invasion of France.[60] With the instability of Sir John French, it should not be surprising that Joffre was able to muscle through his strategic imperatives at the expense of other possibilities, such as the British desire for an attack on Zeebrugge. The expedition to the Dardanelles can in many ways be seen as Britain's first major assertion of their strategic independence from France. Despite Joffre's repeated protestations against any further development of the operations in the Near East the British pressed on.[61] The repeated failures of Joffre to achieve the binding subjugation of the BEF to French orders by having Sir John French become a direct subordinate of GQG were a continued source of strife for the Allies.[62] Friction was not only a result of the British seeking to preserve their independence, but was also sometimes a response to difficult, and contradictory, French demands. In the spring of 1915 GQG was asking that the British get as many new divisions into the field as soon as possible in order to free up French divisions for the attack. At the same time, the French were asking for as many Lewis guns as the British could spare. When the British reported that the Lewis guns were going to be arriving in France behind schedule Joffre became furious

56 Ibid. «*Entrevue avec Lord Kitchener*».

57 Ibid. «*C'est une opération de siège, donc longue et difficile.*»

58 Ibid.

59 Doughty, *Pyrrhic Victory*, pp. 210–211.

60 Doughty, Robert A. (2003), 'French Strategy in 1914: Joffre's Own', *Journal of Military History*, vol. 67, no. 2, pp. 430–432.

61 *SHD*, 16N1905, «Note pour le Général Commandant en Chef (Conversation avec Général Wilson)».

62 Ibid. «*Télégramme Chiffré Confidentiel, Réponse à vos télégrammes 979 et 980*».

and excitedly communicated his anger to the British. In response, the British simply stated that their new units were all going to be outfitted with Lewis guns and that the French would have to choose which they wanted more: fresh British troops or large supplies of British light machine guns.[63]

The situation in which the British were supplying the French with weaponry was a serious role-reversal for 1915. The British were reliant on the French for a wide range of materials – from corrugated iron, to timber, to sand-bags – which the British sourced locally rather than having such items shipped from Britain. British units could only get to the front on French trains, and significant effort went into planning when certain British units could travel from one place to another as their needs were integrated into the larger logistical structure of the French army. One of the major areas in which the British were reliant on French support was artillery, both guns and shells. In Artois, and other sectors, significant help was sought in the form of having French heavy artillery fire on British targets; field artillery batteries were frequently lent to the British outright.[64] In addition to the guns that were lent to the British, either short term for a given offensive or long term for general use, the French also supplied the British with significant quantities of munitions. On 1 May the French ordered the transfer of 52,000 shells to her allies on the Western Front: 42,000 to the BEF and 10,000 to the Belgians.[65] The British hoped that the influx of 10,000 French shells would make the Belgians more amenable to fulfilling a British request for 5,000 shells. The receipt of 5,000 shells from the Belgians would more than double the supply of ammunition that the British had left to serve the guns it had borrowed from 'Little Belgium' (the British then had 2,160 shells to service the 24 guns – just 90 rounds per gun).[66] Thus, some 47,000 shells would be received by the British in preparation for the combined Anglo-French assault on 9 May. To put this number in perspective, the French I CA had roughly 45,000 shells as its store for the First Battle of Champagne only a few months earlier. While this number would prove insufficient to neutralize German trenches, it still represented a significant amount of shells to be parting with in 1915, and was no doubt a factor in the continued strategic subjugation of British desires to French initiatives.

---

63 *SHD*, 16N1905; «La Colonel de la *Panouse Attaché Militaire à l'Ambassade de la République Française en Angleterre à Monsieur le Général Pellé*».

64 National Archives: War Office (WO) 158/26, 'Letter from CGS Robertson to Lt-General Wilson'.

65 *SHD*, 17N303; «Général Commandant *en Chef à Etat-major St. Omer*».

66 *SHD*, 18N134; «Message *Téléphone* État-*major St-Omer à* État-*major Cassel» 26 avril; 19.30.*

Chapter 2
Artillery

There was no more important arm in the First World War than artillery. General Alexandre Percin estimated that over 50 per cent of all casualties were caused by artillery action, thus far outstripping losses due to all other arms.[1] Michel Goya paints a picture of the infantry having been essentially emasculated by the utter domination of artillery on the First World War battlefield. By the end of the war there would be one 75mm gun for every 55 riflemen and one heavy gun for every 70 riflemen; the French army was one of artillery.[2] The success or failure of infantry attacks were all but predetermined by the artillery action on either side with the availability of shells and guns dictating the scale and tempo of operations. Thus, to understand the evolution of French artillery is to understand the evolution of the French army itself.

Artillery underwent a dramatic change through the course of 1915, both materially and tactically. A greater emphasis was placed on infantry–artillery liaison, and the creation of a combined-arms approach, which was conspicuously absent through most of 1914. 1915 first saw what Goya has called the *mathématisation* of combat whereby commanding officers planned offensives based on estimates of the ideal length of front for each infantry division; the number of shells required to kill an enemy soldier or destroy a given length of trench network; and the amount of artillery preparation required to render an enemy position vulnerable to a mass infantry attack.[3] Goya goes on to state that during Second Artois 'the artillery preparation was conducted in a rational and methodical manner for the first time': it would be the first battle fought according to the lessons learned the hard way by front-line troops since the beginning of hostilities.[4] As the first modern and informed artillery battle of the Great War Second Artois would remain of vital importance in the formulation of later doctrine which would guide the French

1 Percin, Général (1921), *Le massacre de notre infanterie 1914–1918* (Paris: n.p.), p. 216. His method for compiling data is not conducive to producing any sort of reliable figure for the entire war and seems slightly jumbled. The figure cited is actually given at 67 per cent but included grenade casualties. Earlier, Percin arrives at the figure of 48 per cent of casualties being attributed to artillery but does so by only looking at those who made it back to the ambulances. At any rate the rough figure of artillery being responsible for between half and two-thirds of all casualties seems thoroughly plausible.

2 Goya, *La chair et l'acier*, pp. 279 and 281.

3 Goya, *Le processus d'évolution tactique de l'armée française de 1871 à 1918*, p. 270.

4 Ibid., p. 271: «La *préparation d'artillerie est conduite pour la première fois de manière rationnelle et méthodique.*»

through the battles of Verdun and the Somme. In order to better understand this evolution it is necessary to start our discussion in the 1890s, a decade which saw the birth of French army artillery doctrine in the writings of Hippolyte Langlois, as well as the birth of modern artillery with the unveiling of the quick-firing 75mm field gun in 1897 – the renowned *soixante-quinze*.

The Guns of August

The French army would enter the First World War with the vast majority of its guns having been designed and manufactured in the late nineteenth century. Certainly the most important of these guns was the 75. Unveiled in 1897, there would be roughly 4,000 75mm guns in the French army by 1914; they were to be the primary weapon for 60 out of the 77 metropolitan regiments of artillery.[5]

With a range of over 6,500m, and a relatively high muzzle-velocity of 500 metres per second, the French 75 was certainly a very good gun, but not necessarily outstanding by the measure of its day. What made the 75 truly revolutionary was its hydraulic recoil system, which absorbed the recoil after each shot by a mechanism which allowed the barrel to slide backwards and then return to its original position without dislocating the entire gun. Before the hydraulic recoil system was invented a gun would be thrown backwards with each shot, and would have to be pulled back into position, reloaded and re-aimed before it could accurately fire again. The 75mm bypassed the majority of this painstaking process, thereby exponentially increasing the speed and accuracy with which a battery could fire. Whereas the de Bange series of artillery (constructed in the late 1870s–early 1880s) could fire two–three rounds per minute, the 75mm could theoretically fire up to 15–20. The 75mm was the pre-eminent field gun in Europe, with 'a battery of four 75s [being] at least as effective as a battery of six German 77s' according to Goya.[6] In 1908 the Russian artilleryman Colonel Belaïeff said that: 'The honour and merit for creating an absolutely new and original rapid-firing field artillery – new material, new firing procedures, new tactical regulations – belong entirely to France.'[7]

As vitally important as the 75 was, it represented only 3,840 out of a total of 12,214 guns available to the French army (31 per cent). The vast majority of the remaining 8,374 guns (8,150) were of the de Bange type manufactured between 1878 and 1882 (Table 2.1).

---

5 Goya, *La chair et l'acier*, p. 150.

6 Ibid., p. 154: «La *batterie de quatre canons de 75 est au moins aussi efficace que la batterie à six pièces de 77.*»

7 Langlois, *L'artillerie de campagne en liaison avec les autres armes*, p. I: «*l'honneur et le mérite de la création d'une artillerie de campagne à tir rapide sur des bases absolument nouvelles et originales – nouvelles matériel, nouveaux procédés du tir, nouvelles règles d'emploi tactiques – appartiennent entièrement à la France.*»

Table 2.1 Number of de Bange guns at the outbreak of war[8]

Number of Guns	Calibre	Maximum Range
500	80mm	7,500m
3,000	90mm	6,000m
1,000	95mm	8,500m
1,620	120L (long barrel)	9,000m
1,500	155L (long barrel)	10,000m
300	155C (short barrel)	6,000m
200	220mm mortar	6,000m
30	270mm mortar	7,000m

As the need for more artillery, especially heavy artillery (*artillerie lourde*, AL), became apparent it was these de Bange pieces – often stripped from the fortresses where they had been stationed for over 30 years – which would provide the immediate stop-gap while production of the modern, quick-firing 105L and 155CTR (*court tir rapide*) Rimailho heavies was underway. Despite their age and painfully slow rate of fire, the de Bange guns remained in active service for much of the war in sizeable numbers, and were often the weapons upon which the success of an attack depended. The guns were rugged, 'virtually brand new' and managed to stand up to the strain and stress of modern warfare with fewer structural issues than the newer and more modern 75mm guns.[9] The 120Ls would be the first French guns in history to be fitted with motorized tractors, and plans were made for these guns (the most numerous heavy artillery piece in the French arsenal) to be fitted with the *affût Mourcet*,[10] a recoil mechanism which would

---

[8] Doughty, *Pyrrhic Victory*, p. 29; Gascouin, Émile (1920), *L'évolution de l'artillerie pendant la Guerre* (Paris: Imprimerie Nationale), p. 29; and Goya, *La chair et l'acier*, pp. 148–150, 162. Unfortunately, it is not certain as to the exact number of guns that the French Army had access to in 1914. The figures above account for 8,150 de Bange guns (a number agreed upon by Gascouin and Doughty once one realizes that Doughty does not include the 120 120L guns attached to the French field army, the 500 80mm guns or the 230 220 and 270mm mortars in his figure of 7,300 de Bange guns), and 3,840 75mm guns. Doughty gives a number of 4,076 minus a supposed 666 reserves, which would make the total number of 75s 4,742, which seems a bit high. Goya gives the total number of 75s as 3,840, which is the figure used here on the basis that Goya cites his number, whereas Doughty does not. There was also the 308 heavy guns the French marched to war with (a figure agreed upon by Goya and Doughty which comprises 84 Bacquet mortars, 120 120L de Bange, which are already accounted for in the above figure of 8,150, and 104 155mm CTR Rimailho guns) and 36 modern 105Ls.

[9] Gascouin, *L'évolution de l'artillerie pendant la guerre*, p. 28: «*presque à l'état de neuf*».

[10] A crude recoil system which could be retro-fitted onto existing guns.

have somewhat increased their rate of fire.[11] Ultimately, the *affût Mourcet* was not seriously pursued. It was decided that such expenditure was useless in light of the many modern 105L guns that had been ordered after the outbreak of war (although it remains possible that some guns did received the *affût Mourcet*, despite the lack of ready evidence to prove it).[12]

Similarly, in an attempt to angle the flat trajectory of the 75mm gun in order to facilitate the bombardment of enemy positions in defilade on reverse slopes, the *Plaquette Malandrin* was invented.[13] The *Plaquette Malandrin* was essentially a set of fins which were added to artillery rounds, which would cause them to fall sharply when fired at medium to short range. In addition to the *Plaquette Malandrin* there was the *cartouche réduite* (reduced cartridge), which also could make shells fired from a 75 fall more sharply – albeit at the expense of shortening the 75's range to 2,000–4,000m.[14] The *Plaquette Malandrin*, a pre-war invention, would be used to fuel arguments that the French did not really need the 220 105L modern heavy guns that had been ordered in April 1913: the same guns whose production would eventually be given as a counter-argument to the production of the *affût Mourcet*.[15] The order would be reduced to a mere 36 guns, which were completed just before the outbreak of war.

The *Plaquette Malandrin, cartouche réduite* and the *affût Mourcet* were three very different methods of solving the same problem: the lack of modern, high-arcing, heavy artillery in the French arsenal. In 1900 the first heavy artillery regiments appeared in the German army and were proudly displayed during their grand manoeuvres that year. The failure of the French to produce their own heavy artillery regiments until 1914 'created a technical gap [between the French and Germans] for which the consequences would be terrible for the French army, which would be obliged to return to using outdated, slow-firing models which excluded … any idea of surprise and audacity in their offensives', according to Goya.[16] The reason for the lack of modern heavy artillery in the French arsenal consists of the complex interaction of the lack of available funds; the poor intellectual standard of many high-level officers (of whom a large number had been appointed based on their political beliefs as much as their merit); and a more general, pervading misconception of how future wars were to be fought.

---

11 Gascouin, *L'évolution de l'artillerie pendant la guerre*, p. 33.

12 Goya, *La chair et l'acier*, p. 161.

13 Doughty, *Pyrrhic Victory*, p. 31.

14 Gascouin, *L'évolution de l'artillerie pendant la guerre*, p. 123.

15 Doughty, *Pyrrhic Victory*, p. 31.

16 Goya, *La chair et l'acier*, p. 155: «Il *s'est donc créé en quelques années un décalage technique dont les conséquences seront terribles pour l'armée française, laquelle sera obligée de revenir à des modèles anciens à tir lent qui excluront … toute idée de surprise et d'audace dans les offensives.*»

The unveiling of Germany's modern heavy guns in 1900 coincides with a general decline in the priority France placed on military preparedness.[17] The years 1901–05 mark the nadir of French expenditure on its armed forces. In these years the French army asked for a total of 357 million francs; they received only 194 million.[18] Of these 194 million francs, the heavy artillery got little, owing to bureaucratic friction and a general scepticism about the efficacy of heavy artillery in the field. As it was, heavy artillery was typically thought of as only being useful during sieges, a form of warfare which was not at all in line with the artillery doctrine of the day, which stressed mobility and speed.[19] Even staunch proponents of heavy artillery often had difficulty answering some of the tough questions on logistics (how does one supply such a large gun with enough ammunition?) and tactics (what is the point of a gun that can fire 10,000m when its crew can only observe up to 4,000m?) levelled at them by their opponents. To make matters worse, the money allotted to the army for new materiel was too often squandered on absurdly complex and costly new mechanisms/programmes which would usually be cancelled soon after they had begun: a problem which underscores the lack of experience and expertise which plagued the upper echelons before Joffre took command in 1911.[20]

Ultimately, by 1910 the French could no longer ignore the growing amount of heavy artillery in their enemies' arsenals; they were forced to invest in modern guns of their own, most notably the 155CTR and the 105L. Unfortunately, only a handful of these modern guns existed in 1914. While the 75 was still a very effective field gun, and would play a crucial role in many early engagements, it was no longer unique by 1914. After 1897 the other powers in Europe had built their own arsenals of quick-firing field guns. Ultimately, the destructive policies of the early 1900s would take France from being the world leader in artillery to lagging behind her main rival, Germany.

Pre-War Doctrine

In discussing the pre-war artillery doctrine of the French army it is probably best to start with Hippolyte Langlois and his 1892 work *L'artillerie de campagne en liaison avec les autres armes* ('Field Artillery in Liaison with Other Arms'). Langlois preached that the two most important qualities of artillery are 'mobility and solidity' (solidity in a material and a moral sense).[21] He especially stressed

---

17 Stevenson, David (1996), *Armaments and the Coming of War: Europe, 1904–1914* (Oxford: Clarendon), p. 54.

18 Goya, *La chair et l'acier*, p. 157.

19 Ripperger, 'The Development of French Artillery for the Offensive, 1890–1914', p. 616.

20 Goya, *La chair et l'acier*, p. 158.

21 Langlois, *L'artillerie de campagne en liaison avec les autres armes*, p. 231.

the 'Frenchness' of mobility and flexibility on the battlefield, harking back to Napoleon. Langlois believed that since artillery could fire in depth, all ranks firing at once, it was exponentially more powerful than infantry, thereby making artillery the principal arm on the battlefield. Furthermore, as artillery can be secretly moved behind one's own lines it can be concentrated at a chosen decisive point unknown to the enemy; achieve a significant local superiority; and fire a concentrated bombardment which would crush the enemy defences in the chosen sector. Langlois placed a great deal of emphasis on artillery quickly blanketing an enemy area with fire in order to catch hidden artillery emplacements, and maintained that '*speed* is the key quality for the attacker'.[22] This notion is further reinforced in Langlois' revised 1908 text in which the speed of the 75, both in firing and manoeuvring, is believed to make his previous theories all the more true.

Langlois was not merely a theoretician; he also concerned himself with the practicalities of field service. His book contains sound, if not simple, equations for ascertaining air resistance, as well as several diagrams which show where batteries should be deployed in various tactical situations.[23] It is important to point out that while Langlois is clearly and consistently in favour of artillery being sited on hilltops firing over open sights, he does present a somewhat nuanced approach. He recognized the threat posed to the artillery by small-arms and counter-battery fire by being exposed, but felt strongly that the speed in correcting one's own fire offset these drawbacks. Langlois also claimed that shells fired from a levelled gun barrel would suffer less air resistance and thereby hit their target with increased velocity. While the exalting of velocity over the importance of keeping one's batteries safe is highly debatable, it is still worth pointing out that Langlois did at least support his conclusions with appropriate calculations.

It should furthermore be noted that there are three 'crests' Langlois refers to: a military crest, on the forward slope, to facilitate rifle fire and visibility; a topographical crest; and an artillery crest, which is actually on the reverse slope.[24] Langlois stated that artillery in this position should be able to fire without fully exposing itself, which certainly shines some light on the oft criticized idea of artillery proudly straddling hilltops, in full view of the enemy and exposed to all manner of heavy and small-arms fire. Above all, Langlois put forward that very simple maxim which is true for all warfare: 'the position of the artillery ought to respond to the tactical goal'.[25]

Perhaps the most pivotal year in the pre-war formulation of doctrine was 1903. In this year Foch published his highly influential *Des Principes de la guerre* ('Principles of War') and the French army produced the *Règlement provisoire de manoeuvre de l'artillerie de campagne* ('Provisional regulations for the

[22] Ripperger, 'The Development of French Artillery for the Offensive, 1890–1914', p. 601.

[23] Langlois, *L'artillerie de campagne en liaison avec les autres armes*, p. 11.

[24] Ibid., pp. 260–261.

[25] Ibid., p. 255: «la position de *l'artillerie doit répondre au but tactique*».

manoeuvre of field artillery'). Together, these two documents laid out a series of important concepts which would form the basis of French artillery doctrine right up to 1914. The first third of the *Règlement* deals exclusively with physical training, with numerous diagrams detailing the correct methods of march and dress, while also providing several pages worth of instructions on gymnastics. Without denigrating the importance of physical training, the amount of time spent on it in the *Règlement* seems wholly out of proportion. This is especially true in light of its rather cursory examination of key topics, from ballistics to tactics.

The *Règlement* officially begins its discussion of artillery by stating that 'speed of fire is the essential property for field artillery'.[26] This popular concept, which obviously played to the strong-points of the French 75, seems to have had a distinctly negative effect on the training of artillery up to the outbreak of war. It fostered a lackadaisical approach to ballistics training, with the French preferring to rely on weight of metal over a judicious use of ammunition to meet stated objectives. This often required batteries to take up exposed positions from which they could, in theory, drown the advancing enemy with fire before being destroyed themselves – a tactic Langlois had advocated. Foch, confident in the offensive power of rapid-fire artillery, supported the use of rapid, blanket barrages in *Les Principes de la guerre*, saying: 'A quarter of an hour's quick fire by mass artillery on a clearly determined objective will generally suffice to break its resistance, or at any rate make it uninhabitable, and therefore uninhabited.'[27] Experience would show sharply just how ineffective 15 minutes of bombardment were to be in the Great War.

After putting forth the need for rapidity of fire, the *Règlement* moves on to discuss range. Over the course of several hypothetical examples of likely scenarios and orders the *Règlement* stresses the firing of the 75 at short–medium range; of the dozen or so examples given the farthest range listed is 4,000m (1,500m short of maximum range).[28] Foch supports the use of the 75 at short range, stating that one is successful when one crushes the opponent '*from a short distance*'.[29] Now, there are certainly compelling reasons to refrain from firing a gun at maximum range. Accuracy is decreased as range increases; firing a gun at maximum range wears the barrel down at a much faster rate than firing at more moderate ranges; and the farther one fires the more difficult it becomes to observe and regulate one's fire – a problem compounded by the *Règlement* stating that observation should be done from the immediate vicinity of the gun. These would all prove to be substantial problems during the war. That said, one should consider the fact that the German 77 had a maximum range of only 5,500m, thus giving the French a

26 FRANCE, *Règlement provisoire de manoeuvre de l'artillerie de campagne*, p. 66: «La *rapidité du tir ... est la propriété essentielle du canon de campagne.*»

27 Arnold, Joseph C. (1978), 'French Tactical Doctrine 1870–1914', *Military Affairs*, vol. 42, no. 2, p. 64.

28 FRANCE, *Règlement provisoire de manoeuvre de l'artillerie de campagne*, pp. 92 and 130–143.

29 Arnold, 'French Tactical Doctrine 1870–1914', p. 64.

1,000m advantage. Even if capitalizing on this disparity may have proven costly in terms of losing guns to over-use and malfunction, the tactical advantages of being able to engage in counter-battery fire with near impunity are immeasurable.

The *Règlement* does eventually begin to introduce mathematical equations regarding the firing of artillery which remained largely relevant, almost certainly due to their simplicity. For example, to find the difference in altitude between a target and one's battery one takes the difference in millimetres as measured on one's *lunette de batterie* (a telescopic viewing device) and multiplies by range, the result equalling the difference in altitude (5/1,000m multiplied by a range of 3,000m equals a difference in altitude of 15m).[30] There are no complex equations for ascertaining trajectory or defilade; nor is any way provided to calculate for range and accuracy modifiers (barrel wear, air resistance etc.). Perhaps the most compelling part of the *Règlement* deals with *tir progressif*, a practice in which an artillery piece would fire two rounds at an increasing range of 100m per pair of shells four times, for a total of eight rounds fired.[31] Not only did this become the standard operating procedure for anti-aircraft fire, but it may also have been a possible source of inspiration in what would come to be known as the 'rolling barrage'.[32] It should be noted, however, that *tir progressif* was banned by GQG during the early phases of the war as it was seen as consuming an inordinate quantity of munitions.[33]

While *Des Principes de la guerre* was primarily concerned with the theory of war, Foch did address some practical issues pertaining to artillery. As with Langlois, Foch recognized the great power of the artillery owing to its range, mobility and the fact that, if positioned well, every gun in an artillery regiment could fire on the enemy simultaneously (a feat impossible for infantry units).[34] Foch envisioned an artillery duel with each side vying for superiority over the other at close range.[35] In an interesting opposition to official works of doctrine, Foch specifically highlighted the artillery's role in preparing enemy positions for an infantry attack, as opposed to merely helping the infantry advance by suppressing the enemy. Foch foresaw the dependence of the infantry on the artillery, and that an increasingly close cooperation between both arms would be necessary in any future conflict.[36] However, he would not go so far as to predict the digging of vast trench networks against which the artillery would have to fire, and also perhaps represented the effect of artillery fire on morale as being more important than it

30 FRANCE, *Règlement provisoire de manoeuvre de l'artillerie de campagne*, p. 109.

31 Ibid., p. 93.

32 Dubois, *L'artillerie de campagne dans la guerre actuelle*, pp. 163–165.

33 Ibid., p. 127.

34 Foch, Maréchal (2008), *Oeuvres complètes, Tome I: Des Principes de la Guerre* (Paris: Economica), p. 413.

35 Ibid.

36 Ibid., p. 414.

actually would be. Nevertheless, Foch's pre-war thinking was not without some valid insight into the nature of modern war.[37]

Only a year after the *Règlement* and *Principes de la guerre* were published the Russo-Japanese War broke out. The fighting in Manchuria from 1904 to 1905 in many ways mirrored the fighting that was to take place on the Western Front during the First World War. French theorists did not always interpret the conduct of contemporary wars in a light which would help prepare the French army for the First World War. The poor performance of Russian artillery, which had been unable to observe and control defilade fire effectively in Manchuria, was seen to support the French doctrine which itself had all but forsaken firing from defilade.[38] Furthermore, the idea that one could be successful by sheer weight of metal was reinforced by the experience of the Russo-Japanese War in which such an unforeseen and inordinate amount of ammunition was expended by both sides.

In 1909 Langlois published *Lessons from Two Recent Wars*, which discussed the tactical and strategic implications of the Russo-Turkish War and the South African (or Boer) War. In this text he set forth a number of ideas that remained strong in French doctrine leading into the war: that the passive defence leads to inescapable failure; the inherent difficulty of a frontal assault; and how the increase in the power of artillery favours the attack. Still, Langlois does point out some very interesting concepts about the increased value of 'hasty field fortifications' and how use of such fortifications would allow the 'assailant [to have] positions as strong as those of the defence'.[39] Langlois developed the idea of using field fortifications further, stating:

> If … we were to construct numerous trenches forming a strong firing line, were to securely protect their flanks and support them in the rear by other trenches, one behind another, we should arrive at a position which would be invulnerable against artillery. This invulnerability would depend not so much on the strength of any one or of the component parts, but on their number and their extension.[40]

After penning this truly profound insight into the current state of warfare Langlois claimed that conceiving a way to *attack* such a fortification successfully is 'beyond the scope of this study', and with that Langlois leaves the problem in the hands of other capable theorists.

By 1913 virtually nothing had changed regarding the tactical conceptions of field artillery set forth 10 years earlier. The 1913 regulation, *Décret du 28 octobre 1913 Portant Règlement sur la Conduite des Grandes Unités (service des armées*

[37] Ibid., p. 415.

[38] Ripperger, 'The Development of French Artillery for the Offensive, 1890–1914', p. 604.

[39] Langlois, Hippolyte (1909), *Lessons from Two Recent Wars: The Russo-Turkish War and South African Wars* (London: Mackie and Co.), p. 138.

[40] Ibid.

en campagne), reinforced old notions of the role of artillery. The *Décret* maintained that 'the artillery has as its essential mission to support the forward movement of the infantry'.[41] Comments on the inherent failure of a passive defence, as well as on the idea of a 'decisive battle' and the role of artillery being to provide a morale boost for the infantry, fall well in line with Foch's notions on the importance of 'mass', 'shock' and impassivity in *Principes*.[42] Another regulation from 1913 states emphatically that 'the essential role of the artillery is to support attacks ... the artillery *does not* prepare attacks, she supports them' [italics mine].[43] It was with these idealistic concepts that the French artillery would enter the First World War.

Baptême de Feu and the Early Trench Battles

With the outbreak of war in 1914 the active French artillery – 3,850 75mm guns and 308 assorted heavies – would be required to overcome myriad doctrinal and material shortcomings before becoming an effective force in the field. The Battle of the Frontiers saw the artillery often left behind by its infantry, who frequently launched impetuous attacks without seeking the support of other arms. The few struggling batteries fortunate enough to eventually fire against the enemy did so 'without lateral liaison, without cohesion'.[44] Still, firing over open sights from hilltops down onto densely massed German infantry allowed the field artillery to act in the role designed for it in the years leading up to the outbreak of war. It is likely that this affirmation of pre-war doctrine retarded tactical adaptation.[45]

Throughout 1914 and into 1915, 75mm cannons were continually used well under their maximum range: firing at great distances was a 'heresy condemned by regulations and by the Command'.[46] Of course, there were several practical reasons why most artillerymen shirked from firing at long range. Terrestrially based observation and fire correction could be quite difficult, especially without the use of forward observation in 1914 and into 1915. Aerial reconnaissance had been used in 1914 to great effect at the army level, furnishing key intelligence about enemy troop movements, but its use for the tactical benefit of artillery was

41 FRANCE, *Décret du 28 octobre 1913 Portant Règlement sur le Conduite des Grandes Unités (service des armées en campagne)* (Paris:,1913), p. 39: «*l'artillerie a pour mission essentielle d'appuyer le mouvement en avant de l'infanterie*».

42 Arnold, 'French Tactical Doctrine 1870–1914', p. 64 and Foch, *Principes de la guerre*, pp. 165, 168, 171 and 176.

43 Gascouin, *L'évolution de l'artillerie pendant la guerre*, p. 52: «le *rôle essentiel de l'artillerie est d'appuyerles attaques ... l'artillerie ne prépare pas les attaques, elle les appuie*».

44 Ibid., pp. 74–76: «sans liaisons *latérales, sans direction d'ensemble*».

45 Rimailho, Émile (1924), *Artillerie de campagne* (Paris: Gauthier-Villars), p. 109.

46 Gascouin, *L'évolution de l'artillerie pendant la guerre*, p. 94: «*hérésie condamnée par le Règlement, et parle Commandement*».

still irregular.[47] Part and parcel of this problem was the fear of friendly fire, which frequently inhibited artillery from taking up a direct infantry support role, and eventually led to the strict timetables of mid- to late war battles.[48]

Furthermore, the rate of fire in some of the more intense early battles (especially the Marne and First Champagne), combined with poor quality ammunition that had been hastily produced in second-rate facilities after the loss of France's industrial North-East, caused widespread and rapid barrel wear. Over-use contributed to a reduction in the range and accuracy of individual guns, while also posing the possibility of barrel explosions and other serious malfunctions. This would be an especially grave problem in First Champagne: after six days of fighting (from 16 to 22 February) exactly 10 per cent, 86 out of 860, of Fourth Army's guns had malfunctioned as a result of over-use.[49] While the advent of interchangeable barrels, in addition to more regular and meticulous barrel oiling, reduced the losses of guns to over-use, it remained an enduring concern throughout the war.[50] Despite methods of reducing the damage done to guns by firing at long ranges, GQG did not officially sanction using the 75 at long ranges until December 1916. At that point the 75mm was authorized to fire at ranges over 5,000m – an act many artillerymen had been performing for years.[51]

* * *

By the end of 1914 the trenches had been dug, forming a continuous line which did not offer any vulnerable flanks for either side to attack, thus bringing the war of movement to a close. Frontal attacks were the only means of combating the enemy on a large scale, and the artillery was forced to adapt to static conditions and to firing against strong enemy field fortifications. Batteries of 75s were pushed as far forward as possible into entrenched positions from which, it was hoped, they would be able to command a wide field of fire deep into the enemy trench network while maintaining good terrestrial observation.[52] To facilitate communication between these forward guns and their commanding officers, batteries of 75s had their meagre telephone wire assignment of 500m augmented with commercial equipment purchased in Paris and Switzerland.[53] Telephonic communication was quickly embraced by units in the field, despite the fact that as recently as 1910 no

---

47 Goya, *Le processus d'évolution tactique de l'armée française de 1871 à 1918*, p. 242.

48 One possible solution was proposed in the September 1915 Champagne battle during which white squares were sewn onto the backs of advancing soldiers so as to provide an approximate range for artillery, theoretically allowing it to provide close support to the infantry: it did not prove particularly effective. House, 'Toward Combined Arms Warfare', p. 21.

49 *AFGG, Tome II, Annex 288.*

50 Campana, *Les progrès de l'artillerie*, p. 43.

51 Gascouin, *L'évolution de l'artillerie pendant la guerre*, p. 92.

52 *AFGG, Tome II*, p. 141.

53 Goya, *Le processus d'évolution tactique de l'armée française de 1871 à 1918*, p. 232.

specific organization for army telephone use had been in existence.[54] The body and base of the 75 was made more secure, and a redesign of the 75mm projectiles allowed for the 75's maximum range to be increased by 25–37 per cent.[55] To make up for the shortage of heavy artillery felt across the front, and in an attempt to encourage better infantry–artillery liaison, Joffre added one organic battery of long-range, heavy guns to each army corps and began to transfer hundreds of de Bange guns from coastal and land-based fortresses to units in the field.[56]

Tactically, the artillery experienced equally large changes coming from both the top down as well as from the bottom up. Artillery quickly adapted to the realities of modern war and began to prepare attacks, instead of merely supporting them.[57] Some batteries had already begun to use artillery barrages to shield attacking infantry and keep enemy reserves from intervening – an important tactic in later battles. Such change was largely unofficial and spread laterally from one unit to another. This process was greatly aided by a static front, which meant that units would be able to maintain lateral liaison and exchange information on new and effective methods. This process was transmitted not only to neighbouring units, but also to new recruits as they were called up via apprentice learning,[58] which the artillery could pursue more effectively than other arms: a result of the lower casualty rate of artillery batteries compared to infantry companies.[59]

While units on the ground were improvising and developing new and effective methodology, Joffre set out to do what he could from above to make French artillery more effective in the field. On 28 November 1914 Joffre officially created the advisory position of 'General Commanding the Artillery', which would coordinate artillery activity within a given corps or army in order to maximize the efficiency of their limited supply of guns.[60] GQG sent out numerous documents to front-line units instructing them to build multiple pre-prepared artillery emplacements in order to facilitate the rapid relocation of batteries behind friendly lines in an attempt to thwart German efforts at counter-battery fire (recall Langlois' assertion that one of the biggest advantages of artillery was its ability to move freely behind friendly lines).[61] GQG also recommended the use of 'dummy' guns and camouflage to hide their batteries from German aerial reconnaissance. Perhaps the most crucial instruction sent down by GQG was one which demanded that batteries maintain extensive logs of their actions, which they could later refer to

54 *SHD*, 19N1677; «Organisation du Service *Téléphonique dans l'Artillerie*» 2 *mai 1915; no. 2917.*

55 Goya, *Le processus d'évolution tactique de l'armée française de 1871 à 1918*, p. 234.

56 *AFGG*, *Tome II*, p. 139.

57 Goya, *Le processus d'évolution tactique de l'armée française de 1871 à 1918*, p. 231.

58 The passing of learning orally from more experienced to less experienced soldiers.

59 Goya, *Le processus d'évolution tactique de l'armée française de 1871 à 1918*, p. 231.

60 *AFGG*, *Tome II*, pp. 139–140.

61 Ibid., pp. 140–141.

in order to quickly register upon a range of targets within their field of fire.[62] More importantly, this information could be relayed back to GQG, thereby connecting the patterns of innovation taking place at the front with those taking place at GQG, which would allow the French army to evolve more quickly and appropriately than before.[63] Eventually, the information pouring in from the front lines would be used in the writing of Note 5779.

Armed with increasing numbers of heavy artillery, and new methods largely gleaned from the experience of combat, the French artillery seemed ready when Joffre, on 30 November 1914, issued a note to all army commanders stating his intention to return to the offensive.[64] Significant efforts were made to ensure the success of Tenth Army's XXI, XXXIII and X CAs in their attack against Carency, Ablain-Saint-Nazaire, Notre Dame de Lorette and Vimy Ridge: the First Battle of Artois. Tenth Army was supplied with six additional heavy artillery batteries, bringing its total number of heavy artillery batteries up from 28 to 34. Tenth Army also received reinforcements of infantry, including 14 elite *chasseurs* battalions, which were distributed between the XXXIII and XXI CAs.[65] The infantry dug saps and trenches up to within 150m of the enemy to shorten the killing zone.[66] Planning for this battle would be of unprecedented depth, with the Tenth Army chief of staff not only providing a detailed plan for the battle, but also offering two well-researched alternatives. These proposed launching attacks in the regions of La Bassée or to the south of Arras, respectively. Within these plans two different operational models were proposed: the battle could be 'progressive' or 'brusque'; it could be a slow battle to gain key points of terrain or it could be a rapid general offensive aimed at breaking the local enemy resistance in one fell swoop. After much deliberation a brusque attack was agreed upon – a far cry from Foch's assertion that 'it is less important to advance rapidly, *but essential to move securely*, step by step, as each objective is gained'.[67]

Despite the seemingly exhaustive measures being taken in the build-up to First Artois, not every commander was confident of success. General Fayolle, then commanding 70e DI (under Pétain's XXXIII CA), wrote on 29 November:

> Conference at Saint-Pol … I ask myself if the senior generals who have been promoted up in this war are not those who do not concern themselves in any way with the human lives which are entrusted to them. Attack! Attack! It's easily

---

[62] Ibid.

[63] Goya, *Le processus d'évolution tactique de l'armée française de 1871 à 1918*, p. 258.

[64] Lucas, *L'Evolution des idées tactiques en France et en Allemagne pendant la guerre de 1914–1918*, p. 1.

[65] *AFGG, Tome II*, pp. 160, 171.

[66] Doughty, *Pyrrhic Victory*, p. 128.

[67] Quoted in Marshall-Cornwall, *Foch as Military Commander*, p. 150.

> said … And when I have captured Carency, at the cost of 500 dead and as many wounded, the question will be: have we advanced even one step? No.[68]

On the next day he wrote, 'I slept poorly because of the discussion yesterday … the sole way to succeed in their eyes is to just kill everyone'; and a few days later, 'It will be a terrible consumption of men to no avail.'[69]

Ultimately, Fayolle's grave concerns manifested themselves in the minds of his superiors, who began to fear that the operation might not be successful. The plan of attack was modified. Instead of one large attack by all three corps simultaneously it was decided that the corps would attack one at a time, over three days, with the AL being shifted around between them. Thus, every heavy gun available to Tenth Army would support XXI CA's attack on 16 December; they would then all focus their attention on supporting X CA's attack on the next day and so on.[70] Sadly this was not enough, and the attacks from 16 to 18 December were a nearly unredeemable failure. Nowhere had the artillery preparation been sufficient, with the blame falling on the gunners who, it was said, had held back for fear of running out of ammunition: a legitimate concern when each field gun was only allotted 400 rounds apiece.[71] The assertion by Tenth Army commander Louis de Maud'huy that he had only had sufficient artillery to attack across a front of 1,500m (roughly speaking, the frontage of a single division) appears to have been correct.

Despite the failure of First Artois, it did foster some positive change. First off, General Pétain had, for the first time, allowed the AL to be commanded by divisional commanders. This decentralization of command helped to ensure a closer artillery–infantry liaison and would be a crucial factor in the effective use of artillery during Second Artois. Furthermore, Joffre released a note stating the vital importance of assuring a successful preparation before the infantry attack regardless of the material costs.

The Artillery Revolution

The period between First and Second Artois saw great advances in the means by which the artillery interacted with other arms, most notably the infantry and

---

68 Fayolle, *Cahiers secrets de la Grande Guerre*, p. 60: «*Conférence à Saint-Pol … Je me demande si les grands chefs qui se mettent en avant dans la guerre actuelle, ne sont pas ceux qui ne se préoccupent en aucun façon des vies humaines qui leur ont confiées. Attaquez! Attaquez! C'est bien vite dit … Et quand j'aurai pris Carency, aux pris de 500 morts et d'autant de blessés, la question aura-t-elle avance d'un pas? Non.*»

69 Ibid., pp. 61–63: «*J'ai mal dormi parce que la discussion d'hier … la seule façon de réussir à leur yeux, c'est de faire tuer du monde … il fera une terrible consommation d'homes sans aucun profit.*»

70 *AFGG, Tome II*, pp. 177–178.

71 Fayolle, *Cahiers secrets de la Grande Guerre*, p. 63.

aircraft. Together these two arms allowed for the artillery to become a much more effective weapon on the battlefield. Their assistance in the prioritization of targets, in addition to the observation and correction of fire, made artillery fire increasingly efficient. This is also a period of intense *mathématisation*, whereby commanders were producing operational plans which would account for exactly how many shells it would take to destroy a given section of wire and how many man-hours it takes to dig a trench. It was during the first half of 1915 that French artillery made its most drastic changes as it adapted to trench warfare. Henceforth, artillery doctrine would largely be comprised of relatively small alterations of the basic principles set forth in Note 5779.

On 16 February 1915 the French Fourth Army launched First Champagne, the first major offensive since First Artois. The cost was tremendous. On the first day I CA fired 33,000 75mm shells, roughly three-quarters of its store, and lost seven out of 142 75s due to barrel wear and subsequent rupture. Despite such expenditure, and an artillery preparation that had lasted for days leading up to the 16th, success was limited to the capture of the first line of enemy trenches. Difficulties arose as the French began their assault on the second line of German trenches. The first had been sited on forward slopes to maximize the visual and firing range of Germany infantry. The second line, by contrast, was sited primarily on either reverse slopes or in dense woods, which were prepared by the Germans to enhance their defensive capabilities by funnelling French attacks into predetermined fields of fire.[72] Trenches sited on the forward slopes of hills were easily observed and bombarded by artillery; trenches on reverse slopes offered neither of these advantages.[73] As a result, they often did not receive an adequate preparation; French infantry in Champagne would be wasted on attacks against intact defensive structures.

After the first three days of combat lateral liaison ceased to exist; regiments often attacked independently of neighbouring units. Command broke down and planning became minimal, with some operation orders being as simple as *même troupes – même objectifs* (same troops – same objectives).[74] Artillery preparations became absurdly brief – sometimes as little as five minutes long – and, most damningly, they took on a largely uniform intensity and duration: this left no doubt in the Germans' minds as to the time and location of French attacks.[75] Maintaining telephonic communication between front-line troops and artillery was all but impossible owing to enemy artillery fire cutting the wires. Other methods of liaison such as flags and flares were not used with any regularity, and most units relied on runners to transmit information between the artillery, command and the troops.

---

72 Liddell Hart Centre for Military Archives (LHCMA), King's College London, Robertson papers 4/1; 'Translation of German III Army Reports of French Tactics, 13 May 1915'.

73 Ibid.

74 *SHD*, 22N163 and 25N87, various documents.

75 Ibid.

Artillery tactics in First Champagne did experience momentary flashes of competence: crude ideas which would later be refined into acceptable artillery practice. The artillery plan for 14e Brigade on 27 February called for a bombardment beginning two hours before the main assault, as opposed to the 30-minute norm. The bombardment would rise to a crescendo just before the infantry assault, and then transform into a barrage advancing by 25-metre bounds until 150 metres had been traversed, at which point the bombardment would enter its third phase: direct infantry support.[76] This plan, a combination of the relatively long artillery preparation with a modified *tir progressif*, is essentially what would later become known as the rolling barrage.[77] As forward thinking as the theoretical plan was, the actual details, in which the crescendo was to last only two minutes, proved to be insufficient.

In terms of materiel, Fourth Army found itself scraping by on very few guns, many of which were old and outdated. One of the principal weapons Fourth Army had at its disposal during First Champagne was the 58mm cannon. Taken from fortresses, where the guns had sat for decades, they were pressed into the field to fill the French army's desperate need for high-arcing heavy guns and mortars. The 58mm could fire either a 16- or a 50-kilogram shell. It had a minimum range of 80m and a maximum of 550m. It was a very difficult piece to aim: a recalibration of 60mm produced a 400m difference in where the shell hit. Even when aimed precisely the piece was still grossly inaccurate. In testing, one 58mm cannon firing eight rounds at a target 400m away managed to land only one within 5 metres of the intended target.[78] To make matters worse, not all of the ammunition was uniform: some 58mm shells had their wings riveted on, others were welded. This was problematic as shells with their fins bolted on had a tendency to lose them in-flight, which meant that fully one-quarter could be expected to land on their side or rear, effectively becoming duds.[79] Its use, along with the early attempt at a rolling barrage noted above, both vividly depict the gap that existed between French thought on how to conduct operations and their material capacity for doing so.

An increase in the use of trench artillery (such as mortars or the 58mm) began to take place in Tenth Army in February–March 1915. On 26 February 1915 General Maud'huy sent a note to his corps commanders asking them for advice on setting up an organized trench artillery (*artillerie de tranchée*, AT) command structure. Maud'huy recognized that 'owing to the development of the front, it appears difficult for one sole officer to exercise command of his trench artillery sections'.[80]

---

[76] *SHD*, 25N87, various documents.

[77] FRANCE, *Règlement provisoire de manoeuvre de l'artillerie de campagne*, p. 93.

[78] *SHD*, 22N163.

[79] Ibid.

[80] SHD, 22N1500; «Le *Général commandant l'artillerie du 21e C.A. Monsieur le Général commandant le 21e C.A.» 10e Armée, 21e CA; 4 mars 1915*: «En raison du *développement du front, il semble difficile qu'un seul Officier puisse exercer efficacement le commandement des sections d'artilleurs bombardiers qui y sont réparties.*»

In response, he suggested that an officer be designated to command the AT and liaise with the infantry. Command would be decentralized to the extent that while each section of AT (16 men manning four guns) would take up their emplacement on orders from the officer in charge of the AT, the section would independently organize its munitions and firing. Maud'huy and his corps commanders estimated that the influx of 38 AT cannons per army corps would require the creation of 160 officers and NCOs to command the guns. It would also necessitate an expansion of the artillery park to handle the increased logistical strain. In sum, it would entail the devotion of some 540 men to trench artillery per corps.[81] This demand for manpower was to be filled by infantrymen who would be working in the vicinity of the new trench guns which it was hoped would increase the guns' effectiveness and adherence to the needs of the infantry.

Between First and Second Artois a significant change in the use of aircraft took place. Squadrons ceased to be solely for the use of army command (which had led to their focus on strategic objectives, such as enemy troops movements), and were largely decentralized down to corps command, where squadrons would become directly attached to groups of artillery. One document, '*Instruction sur l'emploi des avions munis de TSF dans la situation actuelle*' ('Instruction on the use of aeroplanes equipped with wireless radio'), is especially important as it was an amalgamation of all the previous documents released on the subject. This document stated that the two main tasks for aircraft are to observe enemy troops first, and to assist in the registration of artillery second, with use of the on-board radio (TSF) being reserved for urgent information such as the opening of a bombardment by enemy guns.[82] TSF transmissions were received via large receptors which were painfully difficult to move, thereby encouraging batteries to remain fairly static. Each day the corps commander was to allocate a given number of aircraft (a GQG instruction from 2 May 1915 suggests that nearly all aircraft be used in this fashion) to the officer commanding the artillery, who would then distribute that number among his various artillery groups.[83] The artillery groups would then assign targets and objectives for the aircraft, which ensured a more detailed and nuanced approach than might otherwise have been possible without the decentralization of aircraft from army down to artillery group.

Observation and aerial photography yielded important data to batteries and infantry alike. This accumulated data not only assisted batteries when engaging the enemy under poor weather conditions or amidst intense artillery activity, but it was also used by infantry to track the development of German defences over time.[84] Early attempts at using aeroplanes to liaise between the infantry and the

---

[81] Ibid.

[82] *SHD*, 19N1686; «Instruction *sur l'emploi des avions munis de TSF dans la situation actuelle*» X *Armée, 30 avril 1915.*

[83] *SHD*, 19N1677; «*Instruction sur l'aviation de corps d'armée*» au GQG 2 *mai 1915.*

[84] *SHD*, 19N1686; «Note *résultat de l'étude des photographie prises en avion*» 58e *DI, 21e CA, X Armée 1er mai 1915.*

artillery during combat were tried with some success. The paucity of available aircraft, and the difficulties involved in signalling between the three arms, meant that more traditional methods (flags and flares) were popular, despite the ease with which the enemy could impersonate friendly communications – a regularly occurring event.

The French artillery assigned to Tenth Army for Second Artois represented a far cry from its pre-war counterpart. It was not only more numerous (Tenth Army had access to 293 heavy guns, whereas the entire French army had marched to war with only 308), but had also been updated substantially, especially in terms of its doctrine. While not all of these updates were complete, they would lay the foundation for artillery practice for the rest of the war. Liaison between the artillery and other arms, especially infantry and aircraft; increasingly complex firing manoeuvres; new technology and new ways of thinking about the role of artillery on the battlefield continued to be refined, integrated and improved. Much of this improvement, especially in terms of technology and doctrine, was born out of the dismal failures the French army suffered in the first months of the war. Poor liaison, an undeveloped doctrine and a lack of sheer mass ensured the failure of efforts like First Artois and First Champagne. However, the French army learned well from their costly efforts in late 1914/early 1915 and would find themselves ready to attempt the bold offensive that Tenth Army would eventually launch on 9 May 1915.

Chapter 3
Manifest Disarray (9–11 May)

The Second Battle of Artois, which raged from 9 May to 18 June, was the largest and best-supplied trench battle on the Western Front to date. Victor d'Urbal's Tenth Army (initially comprised of 18 infantry divisions, distributed among six corps, armed with 782 field guns and 293 heavy artillery pieces) would attempt to capture two crucial pieces of high ground: Notre Dame de Lorette (NDL) and Vimy Ridge (which the French referred to as *cote* 140). The main thrust eastward against Vimy Ridge (to be made by Pétain's XXXIII CA) would be across a narrow, flat plain against the relatively thin German trench network arrayed between Neuville-Saint-Vaast and Souchez. Here the German defences consisted of two shallow positions: one immediately opposite the French, and a second position atop Vimy Ridge. On either flank the German positions were considerably stronger. To the north, the villages of Ablain-Saint-Nazaire and Carency (in addition to *bois* 125)[1] were strongly held by thick German defences built to funnel attacking *poilus* into predetermined fields of fire. Immediately north of Ablain-Saint-Nazaire lay Notre Dame de Lorette, a hotly contested hill with excellent observation over the Second Artois battlefield. The long struggle for NDL had raged since late 1914, and resulted in a very dense trench network in which French and German positions were, at places, within 35 metres of each other. Just south-east of NDL was the *sucrerie* (a sugar factory) and the village of Souchez, both of which represented formidable defensive positions.

To the south, Neuville-Saint-Vaast and the tiny satellite village aptly named La Targette posed substantial problems. As with the other villages, the Germans had built extensive defences in Neuville, and even connected many of the buildings with underground tunnels. These would not only shelter German soldiers during any French bombardment, but would also provide a means for German troops to retreat quickly and safely from, or reinforce, various positions as necessary. This already daunting town was further protected to the south and south-west by a formidable defensive network known as the *Labyrinthe*. Beyond the *Labyrinthe*, Tenth Army's front began to curve east from where secondary operations would be launched to protect the southern flank of the main assault.

Units in Tenth Army would fight the most sophisticated trench battle to date, using weapons and methodology in a detailed, scientific manner such as had never been seen before.[2] Various incarnations of the rolling barrage would be used by different units, most of them tailoring their own barrage to the specific demands

1 French for 'wood'.

2 Goya, *Le processus d'évolution tactique de l'armée française de 1871 à 1918*, p. 271.

of their front. Infantry units would advance not in simple waves towards distant operational targets, but were given highly detailed roles and objectives so as to maximize their efficiency. Assault troops would break through the enemy defences and capture key points of terrain, while secondary waves would clear out captured trenches of any remaining defenders. Further troops in the rear would begin the process of reorganizing the captured terrain and connecting the old French positions with their new forward trenches. In short, Second Artois was the first truly advanced trench battle of the war, with a great deal of the organization and sophistication that would characterize later engagements. After the minutely detailed preparation and execution of Second Artois (and especially the opening phase in early May) battles on the Western Front would never be the same.

A Tale of Two Divisions: Introduction

What follows is not merely an extended battle narrative; it is also an analysis of the French army as a fighting entity. Even with a relatively small and contained time period such as Second Artois, it would be all but impossible to give a sufficiently detailed account of the trials and tribulations of every French unit that saw battle. It would be even more difficult to try to collate this data into some meaningful form, and provide an adequate analysis thereof. As such, the experiences of two carefully chosen divisions (the 77^{e} and the 34^{e}) will be highlighted throughout the battle narrative, and provide an opportunity to pause and analyse how the war was being fought and experienced by French infantry in 1915. 77e *Division d'Infanterie* (DI) was led by the beloved General Ernest Barbot, a highly competent alpine soldier in command of a division composed of two elite alpine regiments (the 97e and 159e, together forming the 88e Brigade) as well as four *bataillons de chasseurs à pied* (BCPs) organically attached to the 77e from the outbreak of war (normally BCPs were shifted around as needed to give attacks extra force and drive). 77e DI was part of Philippe Pétain's XXXIII CA, and thus constituted part of the main striking force in Tenth Army's bid for Vimy Ridge. 34e DI (composed of the 67e and 68e Brigades) was markedly different. A standard metropolitan division, the 34e would have a supporting role during Second Artois. Stationed on the Tenth Army's right wing, 34^{e} DI would be required to attack a complex and formidable defensive position, but without the same logistical support or marshal excellence of 77^{e} DI.

Looking at these two divisions will provide a 'control' to our examination of French military effectiveness in 1915 by preventing overly optimistic or pessimistic conclusions which might be drawn if one focused on only better or lesser units. Doing so also respects the great diversity that exists in any large organization, especially the military. The analysis of these divisions will provide a platform for examining the rapidly changing technology, organization and tactics which had a dramatic effect on infantry combat in the early twentieth century and which reversed a centuries-old trend towards uniformity and order among European infantries.

The experience of trench warfare showed the combatant nations on the Western Front that the infantryman could no longer be a simple rifleman marching in step with thousands of other soldiers identical in appointments and training, seeking to subdue the enemy by sheer weight of metal and flesh. No longer a cog in the machine, the new model for infantrymen would be the specialist, the expert, the individual. Grenadiers, machine-gunners, flame-thrower operators, signallers, *nettoyeurs*, mortar teams and gas troops all constituted key components in this new dynamic form of infantry combat. During the course of the war small squads of trained specialists advancing on discrete targets under the independent command of a non-commissioned officer (NCO) finally replaced battalions in column smashing through enemy positions by brute force. What had once been an orderly, geometric affair composed of thousands of identical, interchangeable parts became a complex, ever-changing chaos in which small, relatively independent groups of men would be expected to move in a singular, controllable direction with unity of purpose.

With the greater diversification of infantry units, and with lower levels of command gaining access to an increasing amount of firepower (and thus, operational freedom), it is natural to expect a contemporaneous diversification in the way various formations conducted battle. Indeed, every unit in a fighting force, from the squad to the army group, will have a unique methodology and mind-set of war. Methodology here has a variety of meanings. Referring to command, a formation's methodology may be efficient or inefficient, more top-down and centralized or more democratic and devolved. Command may be a dynamic relationship between man and officer built on mutual respect and a concern for the lives of one's fellow soldiers. Conversely, it may be inflicted from above, enforced by punishment and rigid in application regardless of its popularity or lack thereof. Methodology also refers to the tactical and logistical components of making war, such as how one organizes an attack; plans and executes an artillery barrage; organizes one's trench network; ensures liaison between the various arms and the many other activities that infantry formations in the First World War were required to perform. Mind-set is much more narrow, but no less important. A unit's mind-set may be aggressive or it may be passive ('live and let live'). Formations may have high morale or low morale; they may be energetic and hard-working or they may be sluggish and downright lazy, all of which contribute to the individual character of a given unit and will be explored over the course of our examination of the 77^{e} and 34^{e} DIs.

The growing complexity of command and control, which resulted from the vast expansion of the size and scale of warfare from 1914 to 1918, created an increasing number of variables which each corps, division, brigade etc. would have to specify themselves, thus further expanding the diversity of French army practice. Doctrine and orders would be interpreted differently by individual units based on their methodology and mind-set of war, as filtered through the ideas and instincts of their commander. Thus, single doctrinal statements or orders given in the field might result in a remarkably wide variety of practice. To further complicate matters units were not equipped identically. The number and quality of artillery, mortars, grenades and rifles not only varied from unit to unit, but also fluctuated within a formation as resources were shifted from one division to another as needs dictated, and as casualties depleted the capacity of a unit. Supplies of ammunition (of all types), as well as specialist arms such as grenades, flame throwers and gas, were all prioritized and concentrated in carefully chosen divisions, as there was not yet a sufficiently developed logistical infrastructure to provide every division with significant quantities of these specialist arms. This is especially true in 1915 when there was a scarcity of all forms of ammunition. In a conflict such as the First World War, in which technology, in its many forms, was such a critical component of success, the presence or absence of materiel could have a marked impact on the outcome of a given engagement. This in turn could dramatically influence the long-term morale and efficacy of a given formation. Such variability must be taken into account in any serious case study. It is only by embracing diversity within an organization that one can hope to reach any conclusions, however tentative, about just what a given organization was capable of at any one time.

Plans and Preparation

The first plan for what would become Second Artois arose on 24 March 1915.[3] That day Foch submitted his plan to Joffre for approval from *Grand Quartier Général* (GQG). Foch's plan called for an attack which would develop in three distinct stages, with each stage being its own separate attack in time and space: undoubtedly an early conceptualization of operational art. First off would be an attack by XXXIII and XX CA against Berthonval, Carency, Ecurie and Roclincourt (whether or not this would include Neuville is not specifically mentioned). This attack would be followed by a secondary attack by XXI CA on Notre Dame de Lorette, Souchez and *cote* 119, a hillock north-west of Vimy. Finally, there would be a third attack to the south of the XX CA which would, in effect, fix the enemy reserves on that part of the front and protect the flank of the more northern attacks.[4] Foch stressed to his subordinate commanders that they should 'consider these attacks from the point of view of their succession in time', making very clear his idea that what would become Second Artois should be a long battle consisting of several distinct parts as opposed to a single breakthrough battle.[5] Soon after having submitted this to Joffre, Tenth Army commander d'Urbal made it known that he was in agreement with the plan and underscored Foch's main point by saying that the attack would be conducted by 'successive bounds'.[6]

---

3 Daille, *Histoire de la Guerre Mondiale*, p. 112 and *SHD*, 22N1832; «*Projet d'Attaque au Nord d'Arras*» 24 mars 1915, GPN.

4 *SHD*, 22N1832; «*Projet d'Attaque au Nord d'Arras*» 24 mars 1915, GPN.

5 Ibid. «*Si l'on considère ces attaques au point de vue de leurs succession dans le temps.*»

6 Palat, *La grande guerre sur le front occidental*, p. 235: «bonds *successifs*».

Initially there was no assigned date for the attack, only a plan. By 6 April GQG had decided to launch the attack on 1 May, a date agreed upon by British Field Marshal Sir John French, whose troops, positioned immediately to the north of d'Urbal's, would also be taking part in the offensive. It was necessary for the date of the attack to be planned well in advance to allow for the accumulation of men and materiel that would be necessary to launch such a large and ambitious attack. As of 6 April only half the army corps which would take part in Second Artois were stationed in the sector (XXI, XXXIII and X CA), while the remaining three (the IX, XVII and XX CAs) were scattered about the front. IX and XX CA had both been a part of Eighth Army, d'Urbal's previous command, which was renamed the *Détachement d'Armée de Belgique* (DAB) after relinquishing these two corps.[7] These troops were thus not only familiar to d'Urbal but they were also accustomed to fighting alongside troops from other nations, and it was hoped that they would be able to quickly acclimatize themselves to conditions in Artois.

Aside from the time required for each corps to move into the Artois sector they would also require enough time to familiarize themselves with the terrain over which they would be attacking, and to prepare their sector by the digging of additional *boyaux* and *places des armes* (fire-steps). Foch had desired a period of several weeks in which each corps would prepare itself by training, refining its trench network, registering artillery on enemy targets and performing other crucial tasks. In the event, time constraints overwhelmed Tenth Army, and d'Urbal had to compromise this goal by allowing only 12 days for his most recently arrived units to prepare for the attack.[8] As a result, some units (especially the IX and XVII CAs) would not even have time to train in the tenets of Note 5779, and would have to try to execute the new doctrine without having had time to explore its frequently unstated nuances and complexities. The note which officially reduced the anticipated preparatory period from several weeks to less than a fortnight was not released until 19 April, 13 days before the planned attack and an equidistant three days between two major influences upon Second Artois: the release of GQG's comprehensive doctrinal overhaul concerning offensive operations on 16 April and the German gas attack during Second Ypres – also in the *Groupe Provisoire du Nord*'s (GPN's) sector – on 22 April.

From Second Ypres to Second Artois

Tenth Army's preparation for Second Artois was violently interrupted by the German gas attack on 22 April which would come to be known as Second Ypres. The GPN had suffered a devastating attack on that day, raising serious concerns about the prospects of the planned offensive in Artois. Despite the large hole in the

[7] Marshall-Cornwall, *Foch as Military Commander*, p. 157.

[8] SHD, 22N1832; «Instruction *Personnelle et Secret à M.M. les Commandants de Corps d'Armée*» *19 avril 1915*.

Allied lines left by the manic retreat of French troops in the face of German gas on 22 April, Foch remained unwilling to back down from his preparations in Artois, leaving few resources to spare for the fighting around Ypres. This left the British with an exposed flank, which could not be secured despite several lukewarm attempts made by the French to regain the ground they lost on the first day of the battle.[9] In response to the serious threat posed at Ypres Foch ordered d'Urbal to make some units available for transfer northwards. D'Urbal designated the 18e DI (part of Tenth Army's reserve) plus a few additional regiments from the IX CA only recently arrived from the Ypres sector.[10] These units would be held in limbo for a week awaiting orders either to redeploy northward or return to their trenches and resume preparations for the upcoming offensive, thus disrupting an important and already foreshortened process.[11]

The British lost a great deal of confidence in their Gallic allies during Second Ypres, and their public complaints (especially those of Sir John French) gave GQG reason to fear that the British would halt all preparations in Artois in order to focus on events further north. As most of the difficulties arose between Sir John French and Generals Putz and Foch during Second Ypres, GQG was in a position to act as arbiter between the British Expeditionary Force (BEF) and the GPN, thus helping to maintain and foster what feeble attempts at operational cooperation there were between the two supposedly supportive forces in the field. This did not mean, of course, that GQG was free from Sir John's ire.[12] GQG had a strong incentive to foster positive relations as much as possible in the hopes of ensuring the cooperation of the British in the upcoming French attack in Artois. On 5 May Joffre sent to Sir John French what had become a standard letter, thanking Sir John for the continued cooperation of the BEF and asking if he still planned on attacking with d'Urbal's Tenth Army on 7 May (as the plan then stood).[13] Joffre wrote similar letters to Sir John French seemingly every day during the build-up to Second Artois, always humbly expressing his gratitude and very gently asking if the British were still going forward with their plans to launch a supporting attack against Aubers Ridge at the same time Tenth Army made its assault on Vimy Ridge. Joffre was wary of pushing the British too hard after Second Ypres, but also needed to ensure their continued commitment to the Artois offensive. These formal letters were one avenue for doing so. Joffre, of course, was not operating

---

9 *SHD*, 18N134, 'Letter from Field-Marshal Sir John French to General Foch' 30 April 1915.

10 Palat, *La grande guerre sur le front occidental*, p. 234.

11 Ibid., p. 235.

12 Greenhalgh, Elizabeth (2011), *Foch in Command: The Forging of a First World War General* (Cambridge: Cambridge University Press), p. 107.

13 *SHD*, 16N1677; «Le General Commandant en Chef à *Son Excellence Marechal French Commandant en chef des Forces Britanniques*» *5 mai 1915; no. 2930.*

alone. Foch was also working hard to keep the strained relationship between the French and British from rupturing.[14]

No doubt disinclined towards further tactical cooperation with the French, General Douglas Haig, who would command the British troops on 9 May, changed the essence of his offensive. No longer an attack to support the French push on Vimy, Aubers Ridge would be a more independent attack, which simply happened to coincide with the French.[15] This attack would be in two prongs which would advance in a pincer movement towards a central point. This meant that the southern thrust of the attack, instead of liaising with its French neighbours, would advance north-east, away from the offensive of IX CA, which would be advancing south-east. This change of plans came despite the efforts of Foch to, at the very least, build a relationship with Haig in the weeks before Second Artois in the hopes of engendering a closer cooperation between the two allies.[16] On the French side, d'Urbal had lent the British two batteries of 115L heavies, one battery of 105s and a number of 75mm field guns, but had only done so begrudgingly: he more than once advocated reducing the number of guns being lent to the British in favour of extending his attack south of Arras.[17] It appeared that Foch's order that 'the mission of the left of Tenth Army is to assure liaison with the British right, to support and prolong its efforts' would only be given the most superficial execution.[18]

The disruption of Tenth Army's preparation and the slow arrival of additional materiel made it necessary to push the launch date back from 1 May to 7 May.[19] 3–6 May would be used for artillery bombardment which would culminate in an intense fire beginning four hours before the launch of the attack ('H-hour') on the 7th, with H-hour being determined by visibility conditions that morning.[20] In an attempt to assuage the difficulties that the GPN had faced, Joffre approved reinforcing Tenth Army with the battered XVII CA on 27 April; XVII CA would begin to install itself in between XX and X CA two days later on the 29th.[21] Only six weeks earlier XVII CA had been heavily involved in the First Battle of Champagne, where both the 33e and 34e DIs suffered heavy losses.[22] The intervening weeks between First Champagne and Second Artois would prove to be an insufficient amount of time for XVII CA to reconstitute itself effectively.

---

14 Greenhalgh, *Foch in Command*, p. 109.

15 Edmonds, *Military Operations in France and Belgium*, p. 7.

16 Greenhalgh, *Foch in Command*, p. 110.

17 Palat, *La grande guerre sur le front occidental*, p. 234.

18 *AFGG, Tome III*, p. 16: «Assurer la liaison avec la *droite anglaise, l'appuyer, puis la soutenir et la prolongeant, telle est la mission à entrevoir pour la gauche de la Xe Armée.*»

19 Edmonds, *Military Operations in France and Belgium*, p. 5.

20 Palat, *La grande guerre sur le front occidental*, p. 237.

21 Ibid., p. 236.

22 Ibid., pp. 101–112.

As the launch date for Second Artois approached the operational plan began to experience sudden changes. On 1 May d'Urbal officially stated his goal that the upcoming battle be fought in the 'continuous' model, with all the troops at Tenth Army's disposal attacking simultaneously. D'Urbal hoped that the attack would be pursued 'from beginning to end with the greatest vigour and result in a pursuit [of the enemy] undertaken without delay'.[23] This method, while mostly in line with that proposed in Note 5779, was sharply divergent from the original plan of attack as laid out by Foch in late March, which called for several distinct attacks taking place in successive stages over a period of time – a policy d'Urbal had then agreed with.[24] While it is not documented why d'Urbal changed the operational plan from one focusing on a number of smaller, more manageable offensives to one large mass offensive it seems reasonable to suggest that such a move was the result of stress and anxiety as much as anything else. In the days leading up to Second Artois, the largest offensive the French had fought since the Marne, d'Urbal often seemed uncertain, fearful and easily persuaded. His dealings with Maistre were especially telling. D'Urbal's orders and suggestions were frequently countered by Maistre, with the eventual result being D'Urbal giving way to Maistre's point of view.[25] It is also likely, based on how often d'Urbal referenced the document in his communication with his subordinates, that the exhortations of Note 5779 that 'every attack should aim at a breakthrough' had an impact on d'Urbal's operational plan. Undoubtedly, knowledge of the Russian defeat at Gorlice-Tarnów on 2 May helped support the perceived need for a single large effort in Artois. In any case, the weakness of the Russian army remained one of the persistent reasons why Joffre encouraged Tenth Army to continue to attack, even after the initial push stalled and the German defences solidified.[26]

Despite plans which indicated that the attack would be launched on 7 May, d'Urbal, in a letter to Joffre, stated that he felt as though he was ready to launch his attack a day earlier if need be. D'Urbal seemed convinced of his own numerical superiority, and was afraid that Crown Prince Rupprecht (commanding the German Sixth Army opposite d'Urbal's) might receive reinforcements if Tenth Army did not attack soon, thus negating the French numerical advantage. In the same letter in which d'Urbal expressed this concern to Joffre he made sure to point out his belief that Tenth Army still had the benefit of surprise.[27] D'Urbal would probably have been justified in fearing that the Germans would be ready and waiting owing

23 *SHD*, 22N1832; «Instruction *Personnelle et Secrète*» 1er *Mai 1915*: «*du commencement à la fin avec la plus extrême vigueur et complétée par une poursuite entreprise sans délai*».

24 Daille, *Histoire de la Guerre Mondiale*, p. 112; Palat, *La grande guerre sur le front occidental*, p. 235; and *SHD*, 22N1832; «*Projet d'Attaque au Nord d'Arras*» 24 mars 1915, GPN.

25 See *SHD*, 22N1472.

26 Doughty, *Pyrrhic Victory*, p. 161.

27 Palat, *La grande guerre sur le front occidental*, p. 236.

to the massive influx of men and materiel to Tenth Army over the last half of April and the intense, lengthy artillery preparation which would precede Second Artois; whether or not this precluded following Foch's original operational plan is much more doubtful.

In terms of the artillery preparation, d'Urbal would stick with one which involved several days of intense bombardment before launching a general infantry assault. As the attack would rely heavily on slow-firing de Bange guns it was the only way that Tenth Army could hope to make a serious impact on the German trenches. This tactic was more or less exactly what had been called for by Foch in his original plan for the offensive, which stated that 'attacks will be prepared several days before their launch by a slow, methodical and prolonged bombardment – notably from heavy artillery'.[28] General Palat, in his historical analysis the battle, was especially vocal in his disapproval of this plan, which 'did not maintain the element of surprise and, despite its complexity, did not assure a complete artillery preparation'.[29]

On 1 May, six days before the offensive was meant to be launched, d'Urbal sent a memorandum to his corps commanders instructing them on how he felt the artillery preparation should be pursued. The first day was to be spent registering fire on a series of targets; the second day would see the preliminary bombardment of key strong-points in the German first and second lines, with special attention paid to the bombardment of villages, all of which the Germans had transformed into formidable strong-points ; the third day was to be a mere continuation of the second, with the attack finally being launched on the following day.[30] On the day of the attack there would be a three-hour-long bombardment, rising to a crescendo 20 minutes before the infantry went over the top. In the last 40 minutes the field artillery (*artillerie de campagne*, AC) would focus on the first line while the heavy artillery (*artillerie lourde*, AL) would attack the second line, and the *boyaux* linking the two. The AC was expected to bear the brunt of the counter-battery work and a portion of it, mainly 75s, would advance at the same time as the infantry in order to provide better fire support to advanced troops.[31] Each AC gun would have 400 rounds to fire on 9 May; the heavy guns would have 100 rounds apiece.

While the artillery plan set forth by d'Urbal certainly left a great deal up to the initiative and insight of his corps and divisional commanders, very few concrete targets were suggested, it represented a significant improvement over

[28] *SHD*, 22N1832; «*Projet d'Attaque au Nord d'Arras*» 24 mars 1915, GPN: «*Ces attaques seront préparées: pendant plusieurs jours avant leur* début par un *feu – notamment d'artillerie lourde – lent, méthodique et prolongé.*»

[29] Palat, *La grande guerre sur le front occidental*, p. 237: «ne *réservait aucune place à la surprise et que malgré sa complication, il n'assurait pas une complète préparation d'artillerie*».

[30] *SHD*, 22N1832; «Note *sur l'Emploi de l'Artillerie dans l'Attaque*» X *Armée, 1 mai 1915*.

[31] Ibid.

plans he had given to the AL on 24 April. On 28 April, four days after d'Urbal had issued these preliminary orders to the AL, Maistre wrote to the colonel in charge of XXI CA heavy artillery to remind him that the note sent out by d'Urbal was 'very vague' and that the colonel should take it upon himself to modify the plan as he saw fit.[32] Orders to the effect that 'machine guns and dug-outs will be bombarded' required fleshing out, and registration on precise, determined targets needed to be conducted with the goal of achieving the same level of accuracy as field artillery. In order to effectively prioritize the destruction or neutralization of different targets Maistre insisted that the colonel commanding the AL liaise with the different divisions within XXI CA in order to learn from them which enemy defences posed the greatest threat to the success of their offensive.[33] Corps commanders throughout Tenth Army would be responsible for such specification and direction during Second Artois as they would receive little from d'Urbal, who would practically abdicate leadership over the course of Second Artois.

On the morning of 3 May the artillery of Tenth Army officially began its preparatory fire, a somewhat misleading statement since many corps had been engaged in preparatory fire since late April. A good use of aerial reconnaissance in fine weather, combined with minimal counter-battery fire from the Germans, allowed Tenth Army's artillery to mount a highly effective bombardment. Reports from both the XXXIII and XX CA stress the silence of the German guns in response to the heavy French bombardment. According to the 78ᵉ Brigade, part of XX CA's 39ᵉ DI, the German response began to grow slowly as the days of near-constant bombardment ground on. The 78ᵉ experienced concentrated *minenwerfer* fire on 6 May, and even a German patrol, which was repulsed with rifle fire. Neither event caused any French casualties.[34] Officers from GQG watched the bombardment in Artois with interest, reporting on its efficacy – the first concrete test of the doctrine of Note 5779. One witness was especially impressed with the attention to detail exhibited by XXI and XXXIII CAs, not only to the initial preparatory bombardment but also to measures which would ensure that the infantry had effective artillery preparation throughout the course of the attack.[35]

While XX and XXXIII CAs could fire on the German trenches with near impunity, XXI CA was engaged in a desperate struggle on the narrow plateau of Notre Dame de Lorette, which had seen constant fighting since December 1914. The Germans maintained a steady bombardment of XXI CA, and remained active on the ground, with French troops noticing significant reinforcement, accompanied

---

[32] *SHD*, 19N1686; «Le Général Maistre commandant la 21ᵉ *Corps d'Armée à M. le Colonel commandant l'A.L.» 28 avril 1915.*

[33] Ibid.

[34] SHD, 22N1357; «*Compte-rendu de la journée du 5 et de la nuit du 5 au 6» 78ᵉ Brigade, 6 mai 1915* and «*Compte-rendu, Journée du 6 et nuit du 6 et 7» 78ᵉ Brigade, 7 mai 1915.*

[35] *AFGG, Tome III, Annex I*, pp. 201–202.

by the improvement of trench-works, from 3 May onwards.[36] This reinforcement and constant reorganization seriously disrupted and disoriented XXI CA's preparatory bombardment, forcing the AL to halt its preparation in many areas for fear of wasting munitions on unoccupied trenches. Maistre commended this action while at the same time pointing out that the AL must quickly resume its preparation on the newly dug German trenches. It is unclear how well bombarded the German working parties were, especially considering Maistre's order on 4 May that the lighter artillery (especially his modern 37mm trench guns) focus on the German rear and second position as opposed to firing on the German workers.[37]

During the course of Tenth Army's preparatory bombardment, an intense and extensive bombardment was made against the *sucrerie*, near Souchez, in an attempt to confuse the Germans as to the precise time and location of the main attack.[38] Nevertheless, most of the heavy artillery was focused on targets in the enemy rear, especially the roads into and out of the sector, with only a few crucial targets near the front – such as the Cabaret Rouge, Neuville-Saint-Vaast and the *Labyrinthe* – being fired upon by the majority of heavy guns. Unfortunately, many artillery shells fired were duds. Fully 50 per cent of the heavy shells fired in the region of Tilleuls on both 3 and 4 May failed to explode: evidence that the declining quality of French shells would seriously impact the preparatory bombardment.[39]

The bombardment of key areas such as the cemetery of Neuville and La Folie was largely undertaken with 120L de Bange guns, which experienced varying degrees of success. The bombardment of the cemetery was specifically noteworthy for how little it was able to soften enemy defences there.[40] A battery of 95mm guns was also firing on Neuville, but to no avail; it was assumed that the shells had not exploded.[41] Compared to the lack of success achieved by the 120Ls and 95s, one battery of 155CTRs proudly reported that its primary target, the *maison blanche* in Neuville, was completely destroyed along with the surrounding area. This battery then went on to bombard the southern portion of Neuville, as well as auxiliary targets in the German second position; behind the vast majority of the infantry successes in Second Artois one typically finds a battery or two of 155CTRs paving the way.

From 3 May XXI CA, for whom the most complete artillery records have survived, had used its artillery to good effect. A battery of 220mm de Bange

---

36 SHD, 22N1471; «Le Général Maistre Commandant le 21e *Corps D'Armée à M. le Général Commandant l'A.L., le Gal Cdt le 13e DI, le Gal Cdt le 43e DI (à titre des renseignements)» 5 mai 1915.*

37 *SHD*, 22N1471; «Le Général Maistre Commandant le 21e *Corps d'Armée à M.M. le Général Cdt le 13e Division, le Général Cdt le 43e Division, le Général Cdt le 58e Division, le Général Cdt le 92e D.T., le Lt.-Colonel Cdt le Génie, le Chef du 1er Bureau du Q.C. du 21e C.A.» 4 mai 1915.*

38 *SHD*, 19N1685; «*Compte-rendu des opérations du 3 Mai 1915*».

39 Ibid.

40 *SHD*, 19N1685; «*Compte-rendu des opérations du 4 Mai 1915*».

41 Ibid.

mortars was being used against the German positions atop Notre Dame de Lorette as well as against the town of Carency to the south-east.[42] XXI CA field artillery, which had effectively silenced two German batteries on 2 May, continued to pursue their counter-battery fire to good effect, during which time the remainder of the AC bombarded *bois* 125. While aerial reconnaissance had been effectively exploited over both Neuville and Notre Dame de Lorette, its use was still a largely improvisational affair, with methodologies changing and evolving right up to the launch of the attack. On 4 May Maistre issued a note to his squadrons instructing them as to new methods for signalling to the batteries: flares were used by pilots to indicate whether rounds were falling too short or too long and direction was indicated by the plane executing a series of pre-determined manoeuvres in the air; turning right implied that fire should be adjusted to the right and so on.[43]

While the loss of guns to barrel wear remained a serious problem for Tenth Army, the situation was much improved compared to First Champagne. It seems entirely likely that the prescriptions of Note 5779, which called for a drastic reduction in the rate of fire during preparation, had been at least partially responsible for reducing the number of guns lost during the artillery preparation.[44] The large number of dud shells that were being fired by 120Ls in the region of Neuville and the malfunction of a number of 58mm shells whose fins were bolted on (resulting in a large number of shells failing to explode, having landed on their sides or rear after their fins had fallen off mid-flight) were not the only difficulties Tenth Army faced. On 3 May the XXI CA artillery commander, General Cheminon, pointed out that their recent shipment of gunpowder had arrived wrapped only in thin paper which tore easily, dangerously spilling gunpowder in munitions dumps and in the vicinity of the guns themselves.[45] He would go on to describe how difficult these packages would be to maintain in the trenches where mud, damp and enemy fire would pose a constant threat, and suggested that any further resupply of powder be shipped in metal canisters.

In between XXI and XX CAs, Pétain's XXXIII CA – which was to launch the main attack against Vimy Ridge – had been supplied with extra heavy artillery and infantry, including the elite Moroccan Division (DM) and several additional battalions of *chasseurs à pied* (BCP). The infantry of XXXIII CA had been rigorously trained both in the tenets of Note 5779 and in their individual responsibilities in the upcoming attack so that, according to the corps' after-action report, 'each unit had its mission clearly determined. Each man knew his role.'[46]

---

42 Ibid.

43 *SHD*, 19N1686; «*Réglage du Tir par Avions au Moyen Signaux*» 21 CA, 4 *mai 1915*.

44 *SHD*, 19N735; «But *et conditions d'une action offensive d'ensemble*».

45 *SHD*, 22N1500; «*Le Général Cheminon Commandement l'artillerie du 21ᵉ corps d'armée à Monsieur le Général commandant le 21ᵉ corps d'armée; compte-rendu*» 3 *mai 1915*.

46 *SHD*, 22N1832; «*Compte-Rendu Sommaire des Opérations du 33ᵉ Corps d'Armée pendant les journées des 9, 10 et 11 Mai 1915*».

The AL preparation had been 'slow, methodical and prolonged', just as Foch had recommended, and artillery–infantry liaison was well developed. The enemy position opposite XXXIII CA was in shambles by the morning of 9 May. On 8 May, for example, the 2e AL *Groupement*[47] fired upon trenches near Carency, including several machine-gun posts, at the behest of 77e DI, who had communicated to the AL that the preparation was as yet insufficient.[48] This bombardment was carried out by two batteries of 155CTRs who completely destroyed the enemy machine-gun posts in question by 10.00, and were able to resume their scheduled programme of bombardment.

What allowed XXXIII CA to be so successful were not only its training and its large supply of heavy artillery, but also the involvement and insight of its commander, Pétain, who ensured that his batteries were firing efficiently and that the preparation had been as complete as possible before the launch of the attack. Observation and reports from the batteries firing on targets for XXXIII CA were both impeccable, with reports stating exactly how many shells landed on a given target and detailed estimations of damage done provided.[49] A unit as well led and well supplied as XXXIII CA could not help but succeed.

Unfortunately, the French simply did not have sufficient resources (mental or material) to equip every corps as well as it had XXXIII CA. On 4 May Lieutenant-Colonel Gigant, commanding the 1er AL *Groupement*, wrote to the general commanding the artillery for the army to state that they simply did not have enough guns or ammunition to complete their mission; the enemy defences could not be sufficiently damaged to ensure the infantry a good chance at success. Furthermore, it would not be possible for the 1er *Groupement* to provide an adequate barrage to protect the advancing infantry once the attack got underway.[50] Three days later on 7 May, after undoubtedly receiving similar reports to the one by Gigant from units across the whole of the front, General Besse, commanding Tenth Army's artillery, rectified his previous orders to the effect that the AL bombardment planned for the day of the attack should start a full day early.[51] As well intentioned as this order was, the artillery would have its work cut out. Atmospheric conditions had not been conducive to terrestrial observation since 6 May, and flight (whether by balloon or aeroplane) was impossible in the heavy rains which continued to fall.

---

47 *Groupement* being a 'group of groups', a level of organization above the artillery battery and its superior, the artillery group.

48 SHD, 19N1685; «*Compte-rendu des Opérations du 8 Mai 1915» X Armée, 2e Groupement, 2e Groupe*.

49 *SHD*, 19N1685; «*Compte-rendu des Opérations des 24h du 7 Mai 16h au 8 Mai 16h» X Armée, 2e Groupement Réserve d'A.L.*

50 *SHD*, 19N1686; «Le *Liet.-Col. Gigant Cdt le 1er Groupement d'Art. lourde à Monsieur le Général Cdt l'Art. de l'Armée*» 4 *mai 1915*.

51 *SHD*, 19N1686; «Note de Service» Lieutenant-colonel Commandant le 1er *Groupe d'Artillerie Lourde, 7 mai 1915*.

The persistent downpour from 6 to 8 May would compel d'Urbal to delay the attack until 9 May.

In an attempt to strike a balance between an expanded artillery programme and logistical limitations which could not provide the requisite munitions to support it, Besse stated that AL bombardment should start as soon as visibility permitted on 8 May, but that rounds must be fired with the 'greatest precision possible' lest precious shells be wasted by an inaccurate bombardment.[52] Once the AL's objectives were attained it was to effect 'very' slow and irregular fire against various German positions to keep the Germans from repairing any damage done during the bombardment that day.[53] The stress laid on frugality during this phase truly underscores how tight munitions supplies were on the eve of the attack. Besse's stern insistence that the AL should liaise with forward infantry units to determine which targets were the most important, and that bombardment should focus on key defensive points like dug-outs and machine-gun nests as opposed to simply firing at enemy trenches, suggests a radical difference between how XXXIII CA coordinated its artillery fire compared to the other corps in Tenth Army. This is not to suggest that all the other officers in Tenth Army were inadequate – Maistre's command of XXI CA would show time and again that he was one of the most prudent generals to serve in the war (he would go on to command an army group in 1918) – but instead to suggest that a fluctuating combination of poor leadership and/or insufficient materiel across five of the six corps in Tenth Army posed a very serious obstacle to success.

While the artillery of Tenth Army underwent its preparatory fire, the infantry continued to ready itself for the assault. Perhaps the most important task still left to many infantry units was the laying of telephone wires.[54] This would be especially difficult for units like XVII CA which had had a limited amount of time to pursue its preparations, but even units which had been in the sector for some time were behind. On 3 May XX CA, for example, noted that many of its telephone wires were still exposed and could not be counted on once combat began.[55] General Fayolle, commanding 70e DI under Pétain's XXXIII CA, lamented on the 6th that 'yesterday the heavy artillery was already behind in its bombardment. The attack

---

52 Ibid.

53 Ibid.

54 It is interesting to note that on 2 May GQG had released a note reorganizing telephonic communications and giving responsibility for the maintenance and oversight of telephone lines to the engineers. The note explains new techniques; underscores earlier statements that wire should be laid in triplicate whenever possible; and foresees a much greater amount of material being available in the future. Unfortunately, the doctrine and the materials would be too late to impact proceedings in Artois. *SHD*, 19N1677; «*Organisation du Service Téléphonique dans l'Artillerie*» 2 *mai 1915*.

55 *SHD*, 22N1357; «Note *du Service*» *39*[e] *DI, mai 3 1915*.

will not be tomorrow ... maybe it will succeed?'[56] The heavy rains which had interrupted artillery fire from 6 to 8 May would also greatly hinder the infantry's preparations, with only minor works and improvements being made in the three days immediately preceding the general offensive. While these days were not very important in terms of artillery bombardment and infantry preparation they still played a crucial role in the success of the initial assault. By greatly decelerating their preparation the French confused the Germans as to just when the attack was going to be launched; in retrospect Foch noted that 'surprise had been completely realized'.[57]

Not only did these two days of relative inactivity have an unbalancing effect on the Germans, but they also provided the French troops with some much needed rest before the attack, the benefit of which must have been invaluable. Morale was high as 9 May approached; so high that even Fayolle, who had spent the previous few days complaining in his journal that the majority of his troops were fat, ignorant, mediocre, dirty youths (he excludes the elite *chasseurs à pied* from this condemnation), entered a positive note into his diary: 'May 8: Tomorrow we will engage [the enemy] under good conditions, I think. The weather is fine, the ground will be dry, the troops rested and numerous. The artillery preparation has gone far.'[58] It is with this self-assured, optimistic mentality that Tenth Army would go over the top at 10.00 the next morning.

The First Phase – Like Marching in Open Terrain

9 May 1915 was nearly a day that would have lived on for generations in French culture and memory as a great military victory. As it was, 9 May saw the single largest French advance of 1915, untenable as that advance would prove to be in the circumstances. The main attack against Vimy Ridge was to be launched by the 39e, 77e and Moroccan divisions, with supporting attacks from 11e DI (faced with the task of subduing Neuville) and 70e DI (which was to work in conjunction with XXI CA in not only protecting the left flank of the main attack, but also in capturing the plateau of Notre Dame de Lorette). Both the 77e and the Moroccan divisions were arrayed in depth on narrow fronts (roughly 1,200m each, as opposed to 3,500m for 70e DI) to allow a heavier and more sustained forward push.[59] One sole division,

56 Fayolle, *Cahiers secrets de la Grande Guerre*, p. 99: «*Hier l'artillerie lourde* était déjà en *retard pour ses réglages. L'attaque n'aura pas lieu demain ... peut-être réussira-t-elle?*»

57 *AFGG, Tome II, Annex I*, p. 498: «*L'effet de surprise a* été *entièrement réalisé.*»

58 Fayolle, *Cahiers secrets de la Grande Guerre*, p. 100: «*Demain elle s'engagera dans de bonnes conditions, je crois. Le temps est beau, le terrain sera sec, les troupes reposées et nombreuses. La préparation d'artillerie a* été *poussée très loin.*»

59 *SHD*, 22N1832; «*Compte-rendu Sommaire des Opérations du 33e Corps d'Armée pendant les journées des 9, 10 et 11 Mai 1915*».

43e (XXI CA), would launch a moderately successful night attack in an attempt to take the strongly held trenches opposite them by surprise (Figure 3.1).

By 04.30 the assault troops had moved into their most forward positions and by 06.00 the four-hour artillery preparation had begun. After careful registration from 06.00 to 07.00 an intense bombardment of the German first and second positions proceeded until 09.40. From 09.40 to 09.50 the artillery fell silent, only to engage in 10 more minutes of rapid fire from 09.50 to 10.00, at which point the infantry went over the top. After this time the artillery shifted to a fire support role for the infantry, with many batteries firing a rolling barrage. Other batteries (firing on a variety of targets in the German second line) readied themselves to fulfil the order to displace and advance if the infantry attack proved successful and risked outrunning their artillery support. For some units like the 77e and Moroccan divisions, whose rapid advance 'gave the impression of march in open terrain', the order to displace and advance came a mere 10 minutes after H-hour.[60]

The DM had arrived at Berthonval Farm in XXXIII CA's sector on 25 April, and by 27 April had finished disembarking and occupying their new trenches. Upon arrival, the DM, under the command of General Ernest Blondlat, set to work reorganizing its sector, digging additional deep dug-outs for the men and artillery as well as additional *boyaux* to facilitate the 'circulation' of troops.[61] The DM furthermore advanced its forward position to within 200m of the enemy trenches, digging several new *places des armes* in the front line and dug-outs in which to store men and munitions in the final hours before the attack. In addition to this, the DM was thoroughly instructed in the tenets of Note 5779 in preparation for the upcoming attack. Of special importance would be the effective use of specialized troops such as grenadiers, *nettoyeurs* and engineers who would be essential in augmenting the strength of the DM's attack, reorganizing captured trenches and maintaining the impetus of the attack.[62]

On 9 May the DM was ready, well trained and well versed in its terrain and its duties; morale was high. Blondlat had decided that the first few waves should not carry their packs with them, and that the heavier gear should only be taken forward by support troops in the third and fourth waves. The first two waves, referred to as 'shock troops' in the after-action report, were instructed to push forward as fast and as far as possible, bypassing enemy strong-points, safe in the knowledge that the subsequent waves would mop up the Germans left behind and consolidate the captured trenches.[63] Blondlat captured the emotion and spirit of the division, telling them: 'I am your father. You know I love you like my own children. If you want to show that you love me too fight well, and kill as many boches as you

---

[60] Goya, *Le processus d'évolution tactique de l'armée française de 1871 à 1918*, p. 271: «des *ses division donnent l'impression de la marche en terrain libre*».

[61] *SHD*, 24N2908; «Rapport des *Opérations du 9 Mai 1915*».

[62] Ibid.

[63] Ibid.

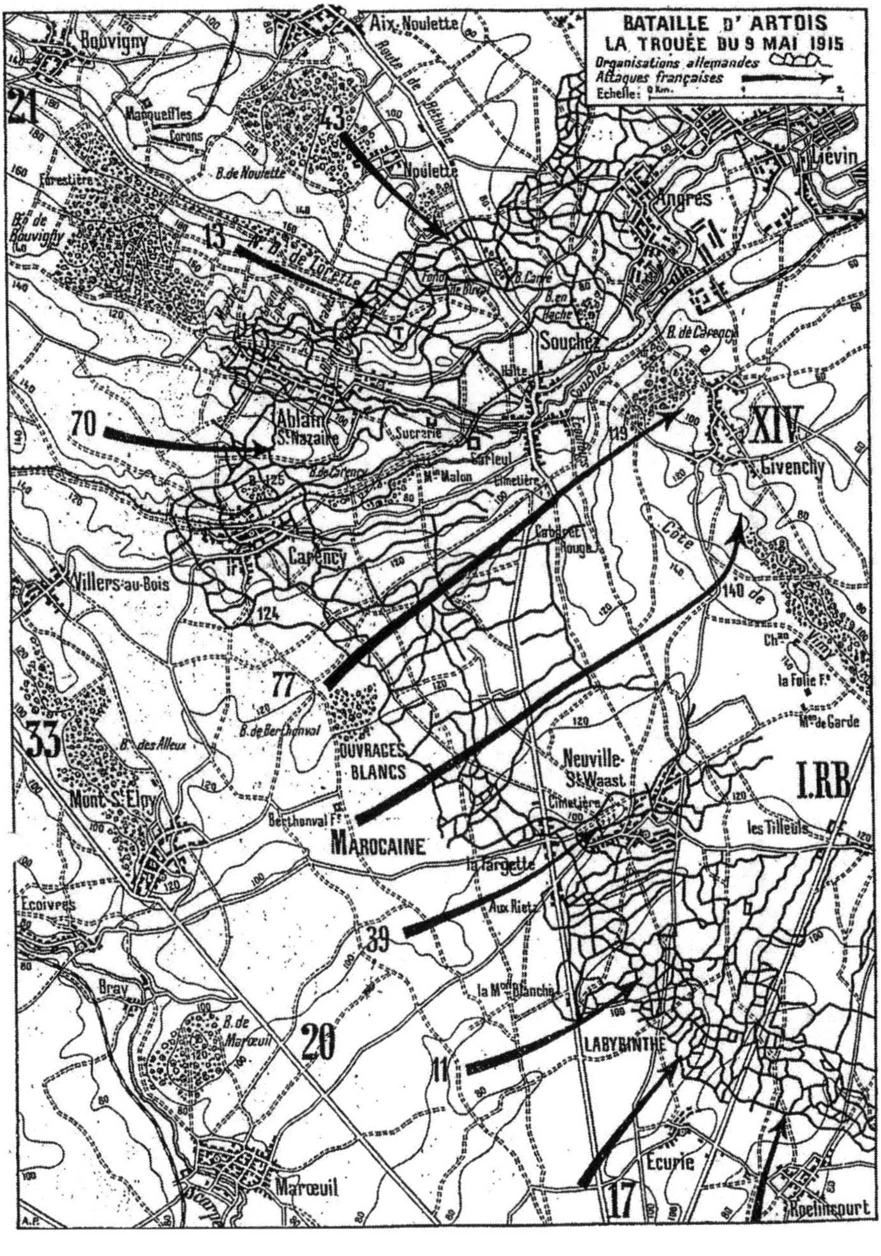

Figure 3.1 Second Artois

Source: Provided by the Musée 1914–1918, Notre Dame de Lorette, Souchez

can.'[64] Between 08.10 and 09.00 Colonel Cros, commanding one of the assault brigades, reported that there were many good-sized holes in the enemy wire.[65] An hour later men of the DM surged through these gaps in the enemy wire, and advanced aggressively towards Vimy Ridge. By 11.30 elements of the DM had reached Vimy Ridge (*cote* 140), and had begun to install themselves in a defensive posture. By noon the number of units on and around *cote* 140 had increased, and by mid-afternoon machine-gun teams would take up positions on Vimy Ridge; they were to play a key role in combatting the incessant German counter-attacks. Some overly aggressive units pushed on as far as Givenchy-en-Gohelle, only to be subjected to friendly fire by a battery of French 75s.[66] At this same time the third wave was busy clearing the first few German lines of dug-outs and machine-gun emplacements left behind by the first two waves. One man of the DM stated that 'the enemy camp was manifest disarray; we met some local resistance here and there but nothing more organized than that was encountered'.[67]

Blondlat had ordered forward a brigade of reserves as early as 10.15; they would arrive at the front two hours later. By 13.30 d'Urbal had learned of the success of the DM and ordered part of 18e DI forward to support the Moroccans; they would not arrive at the front until 17.15. The difficulty faced by these reserves in moving forward under heavy enemy fire, something not faced by the troops in the first few waves of XXXIII CA's attack owing to the speed at which it advanced, would come to be seen by GQG as the primary reason for the ultimate failure to hold Vimy Ridge. Such a conclusion, however, ignores the inherently difficult prospect of maintaining a sharp salient held by isolated troops under violent and continuous enfilade fire. For the DM, progress on 9 May had all but ended by noon, despite a few unsuccessful attempts to capture Vimy Ridge in the afternoon. The rest of the day would be spent in trying to consolidate the position on *cote* 140; fighting off small, localized counter-attacks; and even making a small move on the western limits of Neuville. By 16.00 on 9 May a certain battalion commander from the DM wrote to Blondlat stating his opinion that the DM simply could not hold *cote* 140 with the means at its disposal. The Moroccans had struck deep into the weak centre of the German defences and found themselves isolated and receiving fire from all angles; ultimately they would be forced to retire from Vimy Ridge.

The story of 77e DI is in many ways similar to the DM's. Unlike the DM, 77e DI had been in the Artois sector since before First Artois and was very well acquainted with the area. Its men and officers were highly experienced in fighting

---

64 Ibid. «Je *suis votre Père, vous savez que je vous aime comme mes Enfants, si vous voulez me montrer que vous m'aimez aussi, battez-vous bien et tuez le plus boches que vous pourrez.*»

65 Ibid.

66 Ibid.

67 Daille, *Histoire de la Guerre Mondiale*, p. 115: «Chez *l'ennemi le désarroi* était *manifeste; on rencontrait des résistances locales, mais, plus rien d'organisé ne tenait en face de nous.*»

trench warfare, and the division was led by the competent and beloved General Barbot. By 11.30 on 9 May its 159e Brigade (Bde) had reached Givenchy; the 97e Bde was situated in the cemetery of Souchez; and the 61e BCP was in the Château de Carleul.[68] As with the DM, 77e DI could not maintain such a forward stance for long, and was thus compelled to retire to a line running across the Neuville–Souchez road.[69] Poor intelligence kept 77e DI from having as rapid a response as the DM in terms of reserves being sent forward. What reserves were sent forward moved long after the attack had been launched, and thus were subjected to intense artillery fire.

The story of the rest of the front radiating out from the centre is one of diminishing returns relative to a unit's distance from XXXIII CA. While there is no uniform reason for this, it does conform to what we might consider to be the 'standard' pattern of any First World War battle, whereby the units in the centre advance the furthest while units away from the centre advance to an extent relative to their proximity to the main axis of the attack. Looking to the right of the DM, XX CA (known as the 'Iron Corps'), composed of 39e and 11e DIs, was one of the best corps in the French army. 39e DI had advanced very well on 9 May, capturing the German defensive position at La Targette by 11.15, only to be held up by strong enemy defences in Neuville and the poor advance of 11e DI on their right attacking the *Labyrinthe*.[70] As with the 77e and Moroccan divisions, 39e DI had briefly managed to advance beyond the Neuville cemetery, and had begun to make its move on the southern part of Vimy Ridge before being driven back.[71] Further to the right the XVII and X CAs made very meagre progress, owing to a poor artillery preparation and the large width of no man's land between them and the German trenches opposite.[72] One sombre report from XVII CA stated that 'the losses, which are still not known exactly, are heavy'.[73]

Looking to the left of the main axis, 70e DI progressed acceptably well, considering the immense difficulty of its front. The division faced three occupied villages (Ablain-Saint-Nazaire, Carency and Souchez – all of which had heavily defended cemeteries) and two major defensive points (*bois* 125 and the *sucrerie*), which constituted a very formidable defensive network. 70e DI managed to invest Ablain-Saint-Nazaire, Carency and even the cemetery of Carency, which was

---

68 *SHD*, 22N1832; «*Compte-rendu Sommaire des Opérations de la 77e Division pendant les journées des 9, 10 et 11 Mai*».

69 Ibid.

70 *AFGG, Tome III*, p. 43.

71 *SHD*, 22N1357; «*Résume des Opérations du 9 Mai au 18 Juin 1915» 25 juin 1915, 20 CA*.

72 Daille, *Histoire de la Guerre Mondiale*, p. 116.

73 *AFGG, Tome III, Annex I*, p. 249: «Les *pertes, qui ne sont pas encore connues exactement, sont lourdes.*»

vigorously attacked by the 42e and 44e BCPs.[74] Unfortunately, the poor advance of 360e *régiment d'infanterie* (RI) on the right forced surrounding units to progressively withdraw from such key positions like the Carency–Souchez road in order to maintain a viable front.[75] All told, 70e DI (like the 39e and 11e DIs) advanced roughly 600m.[76] Beyond 70e DI, XXI CA advanced only 200m (not an entirely insignificant advance considering it was made on the very densely held Notre Dame de Lorette plateau), while IX CA advanced even less than that (Figure 3.2).[77]

9 May was a spectacle of fire and steel such as the world had never known. Many infantry units left their assault trenches to find that their artillery preparation had opened up a clear path for them to assault the battered remains of enemy defensive positions. Those units that were protected by a rolling barrage were mostly successful in following them closely enough to be able to pounce on unready defenders, many of whom were found still cowering in their dug-outs. Those formations not blessed with sufficient amounts of heavy artillery and ammunition generally fared much more poorly, and were painfully reminded of the importance of artillery on the First World War battlefield.

Tenth Army's artillery preparation, unprecedented as it was, would raise some serious difficulties for the French army as it honed its tactical craft. In the wake of the stunning advance of XXXIII CA the day before, Joffre met with Foch to discuss the state of Tenth Army's munitions. Out of this meeting came the realization that Tenth Army would clearly have to be more frugal as the attack progressed since GQG could not guarantee that French logistics would be able to keep up with Tenth Army's voracious consumption of shells.[78] This frugality would limit the scale and tempo of operations throughout 1915. Combined with the necessarily slow preparatory bombardments such attacks required, the French army found itself unable to meet its strategic goals with the tactical capabilities at hand, which precluded flexibility.

This point was driven home on 10 May when Joffre sent a letter to Foch detailing a meeting he had had with d'Urbal the day before. At that meeting d'Urbal had proposed an extension of his front south of Arras where he would go on the offensive immediately.[79] Joffre informed Foch of his recommendation that it was inadvisable to launch an unprepared attack against German defences with units that were not familiar with the terrain. By the end of the war offensive means and methodologies would become sufficiently advanced to allow the 'lateral exploitation' here proposed by d'Urbal and long advocated by Foch. In 1915 an

---

74 *SHD*, 22N1832; «*Compte-Rendu Sommaire des Opérations du 33e Corps d'Armée pendant les journées des 9, 10 et 11 Mai 1915*».

75 *SHD*, 24N1754; «*Compte-rendu Journalier du 9 Mai 1915, 24h45; 140e Brigade*».

76 Ibid.

77 *AFGG, Tome III*, p. 41.

78 *SHD*, 16N1677; «*Télégramme Chiffre*» 10 *mai 1915; 21 heures 40; no. 2965.*

79 Ibid. «Le General Commandant en Chef à *M. le Générale Foch*» *10 mai 1915; no. 2964.*

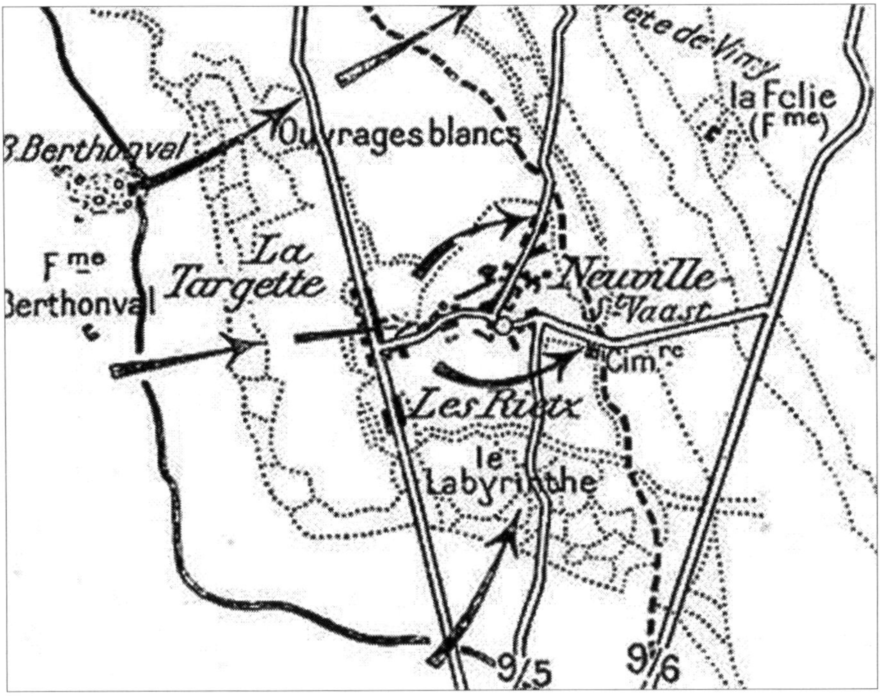

Figure 3.2 The XX CA front

Source: From the 'Illustrated Michelin Guides to the Battle-Fields (1914–1918): Arras, Lens-Douai, and the Battles of Artois' (Clermont-Ferrand, France, Michelin & Cie).

attack such as the one proposed by d'Urbal would require a week of preparation at the very least, thus neutralizing any destabilizing effects that rapidly changing the axis of attack may have had on the Germans. Responding to the logistical situation at hand, and not yet willing to give up on Vimy Ridge, Joffre allocated three fresh DIs for use in Artois in hopes of maintaining the momentum of the French successes the day before.

10 May is a day not often discussed in the historical literature. If the day gets mentioned at all it is often quickly disregarded as a large artillery duel between the French and the Germans during which neither infantry could do much (the activity of a sector still being measured in terms of infantry action by most historians).[80] The French could not renew their attack owing to the excessive disorganization and confusion among the units that had gone over the top the day before; the Germans were still reeling and trying desperately to recover from the shock of the previous

[80] Daille, *Histoire de la Guerre Mondiale*, p. 117 and Palat, *La grande guerre sur le front occidental*, p. 273.

day. The confusion of the French infantry paralysed their artillery, who could not safely or intelligently engage in a renewed preparatory bombardment. Instead, much of the artillery fire was split between bombarding advancing German reserve units and counter-battery fire.[81] Nevertheless, representations which portray 10 May as a day of inactivity do little to explain the importance of this day for operations in Artois. 10 May was a day of attrition during which French troops, wracked with hunger and thirst and cut off from their rear by an impenetrable wall of fire, could not receive reinforcements or resupply in any significant quantity. After a day of agonizing stress, deprivation and fending off local enemy counter-attacks the French assault troops would become increasingly ineffective, thus making a repetition of their stunning 9 May advance all but impossible.

A successful and immediate resumption of the attack was most difficult for formations that found themselves the furthest from their logistics after 9 May; the 77ᵉ and Moroccan divisions, especially, were in no position to attempt a serious continuation of their bid for Vimy Ridge. Both were pinned down in fragile defensive positions and under intense artillery fire which would claim the lives of many of their finest, including 77ᵉ DI's commander, General Barbot, who was killed in his command post at Berthonval Farm on 10 May. A handful of German counter-attacks materialized over the course of the day, including one fairly large-scale attack against the 77ᵉ DI, which were repulsed primarily by rifle-fire.[82] On the wings of this increasingly desperate clinging to the gains of 9 May the 70ᵉ and 39ᵉ DIs did attempt to make progress. 70ᵉ DI would continue its attack on Ablain-Saint-Nazaire and Carency while also supporting the 13ᵉ DI (XXI CA) in its advance along Notre Dame de Lorette. The men of 70ᵉ DI passed the night of 9–10 May digging *boyaux* and consolidating the trenches conquered on 9 May.[83] The troops had been ordered to remain vigilant and to take advantage of any opportunities for artillery and/or infantry action that may present themselves. The plan of attack for 10 May was essentially a flanking movement around Carency in the hope that such a move would force the Germans to pull out of the village. In the event, 70ᵉ DI captured its immediate objectives, made progress in Carency and Ablain, assured its continued liaison with the 77ᵉ and 13ᵉ DIs and ended up on a solid defensive front.[84] 70ᵉ DI, under the impeccable leadership of General Fayolle, would continue to perform with precision and professionalism well into June, long after most other units had lost the ability to make any appreciable forward progress.

On the other side of the 77ᵉ and Moroccan divisions 39ᵉ DI was caught in a very difficult situation. On its left the DM had made spectacular progress, and needed the 39ᵉ to advance as far as it could to provide adequate cover for its right flank.

---

81 *SHD*, 19N1685; «*Compte-rendu des Opérations du 8 Mai 1915*» *X Armée, 2ᵉ Groupement, 2ᵉ Groupe*.

82 *SHD*, 22N1832; «*Compte-rendu Sommaire des Opérations du 33ᵉ Corps d'Armée pendant les journées des 9, 10 et 11 Mai 1915*».

83 *SHD*, 24N1754; «*Ordre pour le nuit du 9 au 10*» *22.45*.

84 Ibid. «*Ordre pour la nuit du 10 au 11 Mai*» *70ᵉ DI, 22.00*.

Conversely, 39e DI was compelled to secure the flank of 11e DI on its right which had made only meagre progress against the densely fortified town of Neuville. Concerned with the dire situation, General Nourrisson (commanding 39e DI) wrote to General Balfourier (commanding XX CA) stating his belief that 39e DI should not be distracted from its mission to capture Vimy Ridge by the problems in and around Neuville; that getting involved in any sort of attack on Neuville would 'immobilize' his division.[85] Nourrisson felt it would be more prudent to surround Neuville and force the Germans to retire via manoeuvre as opposed to wasting infantrymen in difficult and costly assaults on the town itself. This belief was supported by General d'Urbal, who also warned Balfourier that ordering 39e DI to commit its forces to taking Neuville would stunt the further development of the general offensive.[86] Despite these protestations from above and below Balfourier would order 39e DI to turn its efforts southward into Neuville, where the division met with a devastatingly strong German defence which cost the division heavy casualties, and precluded any significant advance.[87]

By 11 May Tenth Army had made sufficient arrangements to attempt yet another general offensive. Very little in the way of resupply or reinforcements had safely arrived to the forward positions of the 77e and Moroccan divisions since 9 May. In sum, the two divisions had received five battalions of reinforcements since 9 May with which to renew their attack on Vimy Ridge. All of these battalions had been subjected to heavy German fire during their long approach march to the front lines.[88] Efforts had been made to enlarge the narrow German *boyaux*, and make them both easier and safer to use, but the work done by 11 May was still largely inadequate. Two days working under heavy fire and threat of counter-attack could not significantly reduce the danger inherent in moving across the shell-swept land between the old French front line and the thin, ad hoc forward position scratched out by the 77e and Moroccan divisions. As a result, information coming to and from the front lines was sparse, and telephonic communication was all but impossible. Despite these difficulties d'Urbal felt it was crucial to try to press home the French advantage sooner rather than later, saying that the attack should proceed 'with the greatest speed , because the more we delay the more the enemy will reorganize his defences and bring against us more men and guns … speed is an indispensable factor of success'.[89]

---

85 *SHD*, 22N1357; «*Résume des Opérations du 9 Mai au 18 Juin 1915*» *25 juin 1915, 20 CA*.

86 *AFGG, Tome III, Annex I*, p. 244.

87 *SHD*, 22N1357; «*Résume des Opérations du 9 Mai au 18 Juin 1915*» *25 juin 1915, 20 CA* and *SHD*, 22N1357; «Lt.-Col. 160e *à Col. Ct 78e Bde*» 11 *mai 1915; 07.00*.

88 *SHD*, 22N1832; «*Compte-rendu Sommaire des Opérations du 33e Corps d'Armée pendant les journées des 9, 10 et 11 Mai 1915*».

89 *AFGG, Tome III*, pp. 47–48: «la plus grand *rapidité, parce que plus nous tarderons et plus l'ennemi se retranchera et amènera contre nous des hommes et des canons … la vitesse est un facteur indispensable du succès.*»

According to the DM's after-action report the situation of the division on 10 May was 'not very pretty, but it wasn't desperate. The losses were enormous but it was necessary to remain in our positions at any cost.'[90] Unfortunately for the men of the DM they were not expected merely to remain in their position, but were counted on to be the spearhead of the attack on 11 May; again the DM, along with the 77e and 39e DIs, would be expected to storm Vimy Ridge. Unlike 9 May, the assault on 11 May would have to be made with men who had not had water in over 24 hours, and had managed very little communication with their commanders or, most importantly, their artillery.[91] Communication was so bad that one unit of Zouaves did not receive its orders until 15 minutes before they were supposed to go over the top.[92]

General Blondlat, who had also been given command of 77e DI for the attack, had planned the two-hour artillery preparation as best he could considering the very limited amount of intelligence available to him. Munitions were at a premium and corps commanders had been advised in a note signed by General Maxime Weygand, Foch's chief of staff, to reserve their AL for attacking long-range targets only, relying on their 75s to take on the brunt of the task of preparing the German position for attack.[93] In an attempt to catch the Germans in the open he had planned for the artillery to increase its range as if the infantry attack was being launched and then return to firing on the German position five minutes later.[94] As the 13.30 H-hour drew nearer the Germans could be seen trying to reinforce their positions opposite the DM, only to be driven off by French machine-gun fire.

When the men of the DM surged out of their trenches they were met with a wall of German artillery and small-arms fire. The DM's position had deteriorated to such an extent by 11 May that they were even subject to flanking fire from the rear pouring out of Souchez.[95] Faced with such odds the DM could not make any noteworthy advances. XXIII CA's after-action report states that 'the Moroccan Division, worn out by three days of violent combat, had lost its aptitude for the offensive'.[96] The same report goes on to say, quite correctly, that the attack on11 May failed owing to its largely improvised nature and that no attack against heavily defended trench-works could succeed without a detailed preparation.[97] In

90 *SHD*, 24N2908; «Rapport des *Opérations du 9 Mai 1915*»: *«La situation n'est pas très belle mais elle n'est pas désespéré, les pertes sont* énormes, *mais il faut que coûte que coûte rester sur nos positions.»*

91 Ibid.

92 Ibid.

93 *SHD*, 22N1808; «NOTE pour M. le *Général Commandant la Xe Armée*» 11 *mai 1915*.

94 *SHD*, 24N2908; «Rapport des *Opérations du 9 Mai 1915*».

95 *AFGG, Tome III*, p. 48.

96 *SHD*, 18N1832; «*Compte-rendu Sommaire des Opérations du 33e Corps d'Armée pendant les journées des 9, 10 et 11 Mai 1915*»: *«La Division Marocaine* épuisée *par trois jours de violents combats a perdu son aptitude offensive.»*

97 Ibid.

the event, the *Division Marocaine*, one of the best divisions in the entire French Army, had been all but wiped out over the course of just three days.[98] The division had lost 43 officers killed, 112 wounded and 11 missing; 508 men were killed, 2,128 wounded and 2,318 were missing, most of them consumed by fire before the crest of Vimy Ridge.

The relief of the DM would begin on the night of 11 May with the 4e and 7e Tirailleurs regiments and the 1e Foreign Legion regiment from 18e DI taking over the right of the DM; the left would have to wait a few more days before being relieved. Even this operation did not go smoothly, with a battalion of the 23e Tirailleurs reporting that they were so exhausted that they could not continue to bring water forward to the front lines and bring wounded men back to the rear. In response, Blondlat had to ask Pétain for a battalion to relieve the exhausted men of the 23e and carry on the crucial resupply and evacuation.[99] 77e DI found itself in a situation not much different from that of the DM on 11 May. Despite spending the night of 10–11 May working furiously to better prepare their trenches and *boyaux*, the captured German position they occupied was still shallow and offered very little protection against German flanking fire.[100] While the losses suffered by 77e DI from 9 to 11 May pale in comparison to those of the DM (77e DI casualties on 11 May were considered 'sensible') very little ground was gained after the first major push on the morning of 9 May.[101]

On either side of the main axis attacks continued to advance slowly through the formidable defences of Neuville in the south and Carency, Ablain and *bois* 125 in the north. 39e DI was still strung out in an L-shape position, with 77e Bde ordered to attack Neuville while 78e Bde was to attack La Folie and Vimy Ridge.[102] 160e RI (77e Bde) had made tentative moves against Neuville on the 10th, but could not advance owing to enemy artillery fire.[103] The German artillery coverage of the area north of Neuville was so complete that any slight movement in that region brought forth a rapid-fire artillery barrage on no man's land and 160e RI's trenches. German artillery dominance in the Neuville area notwithstanding, the German infantrymen exhibited signs of 'nervousness', and let loose undisciplined flurries of panicked rifle-fire several times during the night of 10–11 May. These outbursts were subdued by French artillery.[104]

Despite the losses suffered since 9 May, losses which the commander of the 160e felt were so heavy that they precluded any further offensive action, the regiment would again be ordered to attack Neuville on the 11th. Protestations

---

98 *SHD*, 22N1808; «Division du *Maroc: Régiment de Marche de Zouaves» 11 Mai 1915.*

99 Ibid. «*Télégraphe de Gal Se à* état-*major 33e CA» 12 mai 1915; 07.35.*

100 *SHD*, 22N1832; «*Compte-rendu Sommaire es Opérations de la 77e Division pendant les journées des 9, 10 et 11 Mai*».

101 Ibid.

102 *SHD*, 22N1357; «*Ordre d'Attaque*» 11 *mai 1915; 10.45.*

103 Ibid. «Lt.-Col. 160e *à Col. Ct 78e Bde*» 11 *mai 1915; 07.00.*

104 Ibid. «E.M.B. à E.M.N.» *11 mai 1915, 04.48.*

were sent from the lieutenant-colonel commanding 160e RI to Colonel Lebrun commanding the 77e Bde stating the firm belief held by the officers and men of the 160e that the available French artillery support, despite consisting of a large portion of the heavy artillery in Tenth Army, could not silence the German artillery firing from La Folie and that any units ordered to attack would certainly fail, if not be totally destroyed in the effort.[105] The men of 160e RI had ample reason to be concerned. The artillery preparation which had been planned could not, in any case, be effectively coordinated owing to the cut telephone lines behind 160e RI's position. In an attempt to alleviate the problems stemming from a lack of reliable artillery support it was planned that machine-gun fire would play a vital role in supporting the attack by suppressing German small-arms fire. The inefficacy of this procedure would lead to it being done away with in doctrinal readjustments immediately after the battle.

The concerns of 160e RI would prove to be well founded as the regiment would be 'completely disorganized' (i.e. nearly annihilated) by German artillery and small-arms fire during the attack.[106] The 78e Bde, which attacked La Folie and Vimy Ridge, fared much the same as 160e RI. Four additional infantry companies had been sent to 78e Bde, commanded by Colonel d'Ambly, in the hopes of strengthening that brigade's push on Vimy Ridge. By the time these units had reached the front they had suffered heavy casualties at the hands of German artillery, thus greatly reducing their impact upon the prevailing situation.[107] After the attack of 11 May 78e Bde did not have left any men of 'offensive value', and had lost every single one of its commissioned officers.[108]

On the other side of the 77e and Moroccan divisions 70e DI, working hand in hand with 13e DI (XXI CA), continued its attacks on Carency, Ablain and *bois* 125 while 13e DI moved along the southern edge of Notre Dame de Lorette, advancing towards the Chapel of Notre Dame de Lorette. Conscious of the strength of German machine guns in the area, Fayolle ordered the men to advance in especially dispersed formation as they attacked Carency and *bois* 125.[109] As on the day before, 70e DI met with mixed results; some units, like 226e RI, met with outright failure.[110] Despite this failure to capture the defensive network east of Carency, 70e DI made good progress against the German positions in Carency itself, thus placing the Germans defending Ablain-Saint-Nazaire in an increasingly precarious salient between 70e DI's advances in Carency to their south and 13e DI's advances towards the Chapel of Notre Dame de Lorette in the north. The 13e and 43e divisions of XXI CA would

---

105 Ibid. «Lt.-Col. 160e *à Col. Ct 78e Bde*» 11 *mai 1915; 07.00* and *SHD*, 19N1685; «*Compte-rendu des Operations du 11 Mai 1915*».

106 *SHD*, 22N1357; «E.M.B. à E.M.N.*» 78e Bde*» *11 mai 1915; 19.25.*

107 *SHD*, 22N1357; «Colonel *d'Ambly à Colonel Lebrun*» *11 mai 1915; 18.00.*

108 Ibid. and «Gal 39e *DI à Gal 20e CA*» *11 mai 1915; 20.15.*

109 *SHD*, 24N1754; «*Ordre Général 11e, Attaque du 11 Mai 1915*» *70e DI, 14.30.*

110 Ibid. «La Colonel de *Roig Bourdeville Commandant la 140e Brigade d'Inf.* à *M. le. Gal. C. la 70e Don.*» *11 mai 1915; 06.00.*

ultimately capture the crest of Notre Dame de Lorette (NDL) by the night of 11 May, giving XXI CA a substantial advantage over the Germans atop that crucial high ground. Further to the north IX CA, just like X and XVII CAs to the south of XX CA, did not make any advances to speak of after small, abortive operations merely intended to cover the flanks of the main attack.

As night fell on 11 May the initial push intended to capture Notre Dame de Lorette, Neuville and Vimy Ridge in one sole effort had ground to a halt. After suffering severe casualties, and fighting off innumerable counter-attacks over the course of three days (some small and local, others larger and more organized), XXXIII CA found itself in possession of just less than three-quarters of the ground overrun in the initial attack on 9 May. The foremost units of the 77e, 18e and Moroccan divisions found themselves situated just over a kilometre from the crest of Vimy Ridge. XX CA had made some considerable advances skirting along the northern edge of Neuville, but very little progress into the town itself. XXI CA and 70e DI had made what seemed like only meagre advances each day, but found themselves on the night of 11 May in forward positions which were mutually supporting, well defended and from which further offensive operations could easily be launched.

Pétain was in fact so pleased with the performance of 70e DI, and confident in its continued success, that he pleaded with d'Urbal for permission to launch a medium-scale attack towards Souchez in conjunction with XXI CA on 12 May.[111] After further communication with XXI CA commander Maistre, who informed him that his corps could not engage in such an audacious plan owing to a lack of men and materiel, Pétain gave up the idea in favour of smaller, step-by-step attacks which would take Carency and *bois* 125, Ablain-Saint-Nazaire and Souchez in three successive stages.[112] These attacks, and their analogues in Neuville-Saint-Vaast, would be the shape of operations for the next five weeks in Artois. Meanwhile, Joffre had mediated the transfer of a further group of 75mm guns from the GPN to the BEF, who felt themselves short of artillery in their preparation for an attack on Festubert. In an attempt to pacify Foch, and make the transfer of guns northward more palatable, Joffre reminded him that with the arrival of III CA Foch would be receiving a substantial influx of artillery – more than he could realistically bring to bear in Artois, Joffre erroneously suggested.[113] Faced with the great fatigue and disorganization of its troops, heavy losses and the need to support (and thus compel) the British to take a more active role in the battle, Tenth Army would be unable to launch another general offensive until 16 June.

Despite its failure to capture Vimy Ridge, Tenth Army (and XXXIII CA in particular) had much to be proud of. Through good leadership, ample material

---

[111] *SHD*, 22N1832; «Le Général Pétain commandant le 33e *C.A. à Monsieur le Général commandant le 10e Armée*» 11 *mai 1915; 18.30.*

[112] Ibid. and *12 mai 1915; 17.00.*

[113] *SHD*, 16N1677; «Le Général Commandant *en Chef, à M. le Général Foch*» *12 mai 1915; no. 2983.*

support and the use of a variety of new and daring tactics Tenth Army had succeeded in making the largest single advance against an established trench network to date. Those units which did make substantial progress (namely the 77e and Moroccan divisions) are proof of the potential which the French army's new tactical system provided. Even units that did not fare as well were typically operating at a very high level of competency. Their inability to produce anything like the advance made along the main axis of the attack contrasts the potential of the French army in spring 1915 with some of its limitations, material and professional (the nuances of which will be explored in the next chapter).

Just as during First Artois, operational plans were drawn up on a scale and tempo that was ultimately too grand for the available materiel to support adequately. It was not a problem of the French having insufficient guns or shells; it was a problem of the French planning beyond their means. These limited means were part of the reason Foch had advocated a three-step assault on Vimy Ridge; they would also be part of the reason why Tenth Army's general attack on 9 May ultimately failed to meet its ambitious targets. After 9 May the likelihood of Tenth Army meeting its operational objectives fell precipitously. The Official History states that 'all the attacks [were] going to fail; our troops [were] tired, the units [were] all mixed together; furthermore, the enemy had brought new reinforcements from Lens and Douai'.[114] Combined with the great difficulty of trying to plan and execute fresh artillery preparations on the fly, it is clear that Tenth Army was in no position to secure Vimy Ridge.

The failure of XXXIII CA to maintain its hold on Vimy Ridge owing to the German artillery response and flanking fire from Souchez and Neuville goes a long way to validate the slower, more methodical approach suggested by Foch in his initial planning of the offensive; d'Urbal's brusque plan of attack made success significantly more elusive. Faced with the impossibility of making another immediate attack on Vimy Ridge Foch's methodology would be forced upon Tenth Army as it fought for important pieces of terrain one by one until it was able to launch a second major attack against Vimy Ridge. The continued operations of Tenth Army were made especially difficult as units could not rely on a basic standard of leadership and expertise owing to a general lack of experience with trench warfare, and with the new offensive methods being used. Only time and repetition would build up a sufficient body of experience to even partially regularize how units conducted themselves, and allow for general 'best practice' methods to become ubiquitous. As such, there persisted a wide range of proficiency in Tenth Army, and across all the French forces. This variability posed serious problems for the French high command, and continues to pose problems for historians in their assessment of the French war effort. The following chapter's case study will help to refine our picture of 1915 by providing an in-depth narrative and analysis of the

---

114 *AFGG, Tome III*, p. 47: «*toutes les attaques vont* échouer*; nos troupes sont fatiguées, les unités mélangées; de plus, l'ennemi a fait venir de Lens et de Douai de nouveaux renforts.*»

workings of two divisions in Tenth Army during Second Artois: one successful, one rather less so. By presenting a breadth of experience some conclusions may be reached about the actual effectiveness and potential of the French army, its weapons, its training and its men, in 1915.

Chapter 4
A Tale of Two Divisions, Part I

In an organization comprising millions of individuals it is inevitable that there will be a wide variety of experience. When attempting to recreate the history of even discrete events, like a single battle, it can be incredibly difficult to render this great mosaic succinctly while maintaining a fair representation of the historical data at hand. Efforts to present a full and rich picture of the past are a crucial component of all historical research, but it is especially important when trying to assess the military effectiveness of a fighting force. It is inevitable that certain units will be better trained, led and equipped than others in a modern, industrialized war. As a result, some units will tend to perform better than others, which makes it very difficult to ascertain the tactical effectiveness of entire armies. This chapter, and its continuation in Chapter 7, will attempt to address this difficulty by looking at two divisions which fought in Second Artois: the 77^e^ and 34^e^ *Divisions d'Infanterie* (DIs). The division being the tactical unit in 1915, a review of the experiences of one very effective division (77^e^) and one largely ineffective division (34^e^) will, at the very least, provide a range of experience from which to draw conclusions about the value of the French military system in 1915. These chapters will furthermore permit a closer examination of infantry life and procedures during Second Artois such as more standard battle narratives may not be able to provide, thus further enriching the historiography of Second Artois, and the French army in general.

La Division Barbot

77e DI was one of the elite formations in the French army. Composed of highly motivated, elite soldiers (alpine troops and *chasseurs à pied*), it was the logical choice to lead the attack on Givenchy-en-Gohelle and Vimy Ridge in the opening phase of Second Artois. Its commander, General Ernest Barbot, had moulded a dependable formation; his thoroughness and compassion for his troops is remembered still. Barbot is immortalized in a life-sized statue as part of a memorial to the deeds of 77^e^ DI during Second Artois, which stands in the middle of Souchez. In many ways this was an unlikely end for Barbot. He was 59 years old when he led the attack on 9 May, very old for a divisional commander. His age reflected the slow advance of his career. Having been commissioned in 1877, Barbot had risen to the rank of colonel (commanding the alpine 159e *régiment d'infanterie*, RI) at age 57. Like Pétain, he was nearing retirement when the war broke out; the loss of both his wife and son in 1912 had pushed Barbot to retain his command, and to devote his life to his *alpins*. During mobilization Barbot's 159e RI would be expanded to include 97e RI and four battalions of *chasseurs à pied* (the 54e, 57e, 60e and 61e), thus forming *La Division Barbot*. The division would make a name for itself during its heroic stand before Arras in October 1914. From then on, *La Division Barbot* would be 77e DI, and its commander became known as the *Sauveur d'Arras*: the Saviour of Arras. Unofficially, the division retained Barbot's name, even after his death, and entitled their published divisional history *La Division Barbot*.

Written by Captain Humbert, *La Division Barbot* is more a memorialization than it is a proper work of history. It opens with a short dedication: 'This book is written for the *alpins*, *chasseurs*, artillerymen, cavalrymen, sappers, and the men of all the services of 77e DI.'[1] The style throughout the book remains flowery and romantic, at times even clichéd. While it is not a particularly good source if one is looking for solid historical facts about 77e DI, it does present a good picture of how the men of 77e DI saw themselves and their leader. Barbot is unquestionably the hero of the story, exhibiting unbounded bravery and wisdom at every turn. When describing 77e DI's combat before Arras in October Barbot is described as wearing the 'blue great-coat of a simple soldier of the 159e [RI], wearing a beret flat on his head, hands in his pockets' – every bit the stoic, humble leader.[2] Upon learning that, while he was out, one of his subordinates had given the order to retreat in the face of overwhelming German numerical superiority, Barbot is quoted as having responded: 'An order to retreat? Never! I am still living, we will not retire! And now, *messieurs* staff officers, go! Go to the troops! Lift up their hearts! Repeat everywhere that we are holding out well and that we will not lose Arras!'[3] While it is unlikely that these are the precise words of Barbot, they strongly attest to the impression Barbot made on his soldiers as a leader of men. It was this leadership which would set the stage for the high performance of 77e DI in the opening phases of Second Artois.

On 1 May 77e DI issued its final op-order for the attack which would take place on 9 May. The *alpins* were to attack towards Givenchy-en-Gohelle and *cote* 119, a lone hill slightly north-west of Vimy Ridge, while the *chasseurs* would attack the Château Carleul. The infantry were ordered to advance without concerning themselves with what was happening on either flank or behind them. They were to trust their neighbouring units, and the subsequent waves of *nettoyeurs*, to ensure the continued security of the division's flanks as it marched forward through the German defences.[4] A series of discrete objectives were assigned to each of the various waves of infantry, and each battalion was made fully aware of its role in the wider attack. The assault troops were to leave all nonessential gear behind,

1 Humbert, Capitaine (1919), *La Division Barbot* (Paris: Librairie Hachette et Cie).

2 Ibid., p. 23: «Barbot ... *dans sa capote bleue du simple soldat du 159, coiffé d'un béret posé à plat sur sa* tête, les mains *dans ses poches.*»

3 Ibid. «Un *ordre de retraite? Jamais! ... Moi vivant, on ne recule pas! ... Et maintenant, messieurs les Officiers d'état-major, allez! Allez à la troupe!* Élevez *leurs cœurs! Répétez partout que l'on tient bon et que nous ne lâcherons pas Arras!*»

4 *SHD*, 25N172; «*Ordre d'attaque pour la journée du [blank] Mai*».

taking forward with them only ammunition, three days' worth of food and water, half the division's machine guns plus barbed wire and sandbags to organize an effective defence once they had reached their objectives.[5] Assault battalions were separated into two waves, the second wave following only 100m behind the first, with further reserve battalions following closely behind. Each company carried with it 100 British Mills bombs, which were concentrated in the hands of specialist grenadiers. These grenadiers would also carry forward additional explosives (mostly petards) for dealing with enemy wire or other obstructions.[6] A small number of engineers were also sent over with the assault wave to help organize the forward defence and to provide additional detonating power and expertise for dealing with any particularly stiff centres of resistance.

Likewise, the artillery bombardment was laid out in great detail (although not in final form until 5 May). On the day of the assault the division's half-battery of 220mm mortars would fire on the south-east portion of Souchez; the 155CTRs would focus on the German second line; and the 155Ls would fire on further-off targets, including the enemy observation posts on *cote* 119, the Cabaret Rouge and in between Souchez and Givenchy-en-Gohelle.[7] The 37mm and 58mm guns would fire on strong-points in the German first and second lines until five minutes before H-hour, at which point the crews would begin to ready their guns for relocation forward to the soon-to-be-captured German trenches. The 95mm and 105mm guns were to engage in counter-battery fire. At the hour of attack the artillery firing on the enemy first line would lift to the second line and begin firing a barrage which was to roll forward for 10 minutes, at which point it was to have reached *ouvrage* 123/*route de Bethune*.[8] There the artillery would hold the barrage until signals from the infantry were received (either a telephone call or flares, although the infantry were reminded to use flags if neither of these options proved viable). Once the infantry had indicated that they had reached *ouvrage* 123 the barrage would continue to roll forward until the tertiary objectives of the division were reached. The artillery was further instructed to prepare for the possibility that the infantry would not advance as quickly as planned, thus compelling the artillery to roll back the barrage and fire on any targets that were holding up the advance.

The details of the plan thus enumerated, 77^e^ DI would spend the days before the attack maximizing the utility of their position and working to ensure that the attack would unfold smoothly. On 4 May 97^e^ and 159^e^ RIs were instructed to create extensive depots in their front line sufficient for each battalion to store 25,000 rifle rounds and 1,000 litres of water.[9] The infantry was ordered to take entrenching tools from the rear depots to make up any deficit that may develop in the days before the

---

5 Ibid.

6 Ibid.

7 *SHD*, 25N172; «*Emploi et rôle de l'Artillerie dans la zone de la DIVISION*» *5 mai 1915*.

8 Ibid.

9 *SHD*, 25N172; «88e *Bde à 97e et 159e*» 4 *mai 1915*.

attack, while the grenadiers and *nettoyeurs* were issued with revolvers and trench knives for the close-range fighting which lay ahead.[10] Extra telephone squads[11] were sent to the front to ensure liaison with the rear; troops were issued flares for communicating with the artillery; and front-line commanders were reminded to destroy any sensitive documents before the attack lest they fall into German hands.

In the days immediately before 9 May a veritable barrage of specifications and rectifications were issued by Barbot to ensure that his division was as prepared and informed as it could possibly be. Individual *boyaux* were assigned specific roles in the coming offensive in the hope of maintaining a relatively free circulation of men and materiel once the battle began. Some were to be used only for the forward movement of reinforcements; some were reserved for supplies being moved forward; while others were dedicated for the evacuation of the wounded.[12] Every *boyau* was to have two soldiers at each end to direct traffic, and keep soldiers from using the wrong ones. Colonel Wurster, who would later command 88ᵉ Bde (77ᵉ DI), was to be in charge of maintaining sufficient order to ensure the smooth transit of soldiers and supplies to and from the front, despite the confusion of battle. Companies were ordered to attack with no more than 180 men (out of roughly 250), with the rest staying behind to organize the sector, to stand by ready to resupply the front line, and to begin the work of advancing *boyaux* into no man's land to connect the old French position with the newly captured one. Barbot even went so far in his detailed preparation for the battle as to specify that those wounded in the abdomen were to receive priority for evacuation.[13]

General Barbot also paid special attention to the tactics his division would employ on 9 May. In a note issued on 6 May he reminded his subordinates that the artillery would be firing at ranges of 3,000–4,000m, which meant that the barrage would have a margin of error of 60–80m, and that attacking infantry should be careful not to follow the barrage too closely for fear of friendly fire.[14] Conversely, great stress was laid on the fact that once the barrage passed a German position the Germans would scramble to set up a defence, and the attacking French troops needed to beat them to the parapet or risk being engaged by organized enemy resistance.[15] Barbot went on to reiterate the importance of each man knowing his role in the upcoming attack, as well as pointing out the importance of apprentice training by reminding veterans to take newer soldiers under their wing and instruct them on what things were going to be like on the day of the 'big push'. *La Division Barbot* would be thoroughly and meticulously prepared for battle by the time it went over the top in the opening phase of Second Artois (Figure 4.1).

---

[10] Ibid.

[11] Each telephone squad consisted of two NCOs, two telephone specialists and 2,500m of telephone wire.

[12] *SHD*, 25N172; «Annexe à *l'ordre d'attaque No. 7358/s» 77ᵉ Division, 5 mai 1915.*

[13] Ibid.

[14] *SHD*, 25N172; «*77ᵉ Div. A 88ᵉ Bde, groupe chasseurs» 6 mai 1915.*

[15] Ibid.

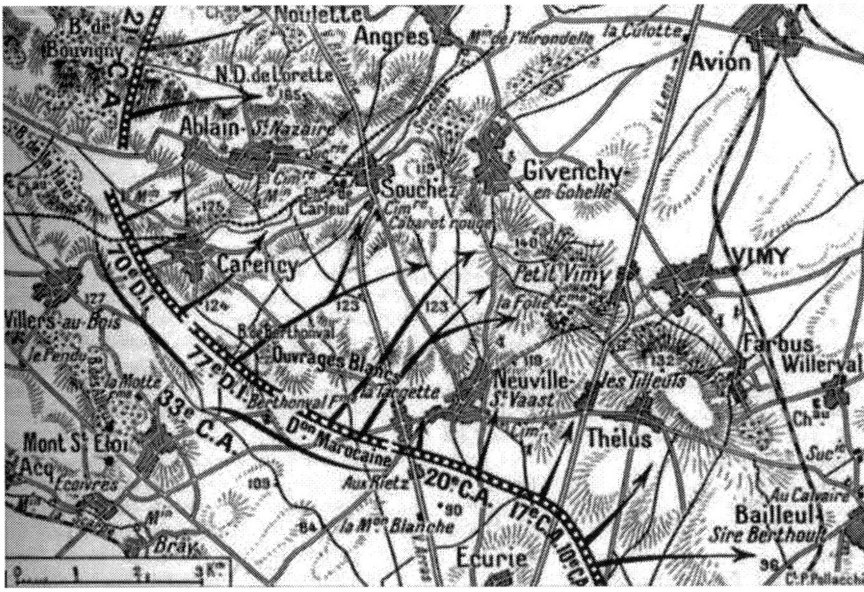

Figure 4.1 The main axis of attack for Second Artois

Source: From the 'Illustrated Michelin Guides to the Battle-Fields (1914–1918): Arras, Lens-Douai, and the Battles of Artois' (Clermont-Ferrand, France, Michelin & Cie).

At 08.00 on 9 May, two hours into the four-hour bombardment which would immediately precede the attack, General Barbot sat in his command post at Berthonval Farm (some 300m from 77e DI's *parallèles de départ*) with his liaison officers, ready to respond to any situation that may arise. Two hours later, at 10.00, the men of 77e DI surged towards Vimy Ridge, Givenchy-en-Gohelle and *cote* 119. The first two German trench lines were passed in a single bound, with the 77e suffering only very light casualties.[16] The infantry had been able to keep close to the barrage, and often reached German positions to find the defenders still dazed in their dug-outs and easily dealt with by French grenades. By 10.40 159e RI was approaching *ouvrage* 123; it was here that the German defence first began to materialize.[17] The Germans in *ouvrage* 123 had had sufficient time to leave their dug-outs and mount a defence, which ultimately had to be broken by a bayonet charge. After subduing *ouvrage* 123, 77e DI continued its advance following closely behind the rolling barrage, and by mid-day found itself fighting atop Vimy Ridge and in Givenchy-en-Gohelle. While the barrage had done a good job of keeping enemy infantry from being able to organize any defence strong enough to stop the division, its slow pace left the infantry exposed to intense enfilade fire as

16 *SHD*, 26N701/1; 159e RI *Journal des Marches et Opérations* (JMO).

17 Ibid.

they crossed the Neuville–Souchez road.[18] German 77s and 105s, plus small-arms fire from *cote* 119 and the eastern reaches of Souchez, inflicted heavy casualties on the increasingly fatigued infantry.

By 14.00 strong German fire and several counter-attacks had forced the 77e DI to retire from many of its forward positions; the division was strung out in front of Souchez, *cote* 119 and Vimy Ridge, straddling the Neuville–Souchez road. The *chasseurs* had been pushed out of the Château Carleul earlier that morning, and had retired to the Carency–Souchez road, where they held on to the left of 97e RI.[19] Initial reinforcements had remained very close to the assault troops as planned; the first reserve battalions arrived at the Neuville–Souchez road only 10 minutes after the assault wave.[20] Here the division would consolidate its gains with an eye to renewing the attack as soon as possible. Tentative assaults on Vimy and Givenchy-en-Gohelle, launched that afternoon in liaison with the Moroccan Division (DM), were fleetingly successful (the divisions were able to capture and destroy several enemy guns) but ultimately resulted in the *poilus* being forced to pull back once again under intense enfilade fire from Souchez and Neuville.[21] A renewed attack on Souchez, Vimy Ridge and Givenchy-en-Gohelle was planned to occur at 18.00. However, the impossibility of liaising with the artillery and coordinating an effective bombardment forced Barbot to call off the attack and focus on further consolidating his forward positions.

On 9 May 77e DI had advanced over 4 kilometres before being pushed back, and had taken 600 prisoners (20 officers, 580 men), six German 105mm heavies, two 210mm mortars and several machine guns.[22] 77e DI lost an estimated 2,500 men and 60 officers in the effort.[23] Yet, the division had failed to capture *cote* 119 and Givenchy-en-Gohelle, much less secure Vimy Ridge. Resumption of the attack was ordered for 09.00 on 10 May, but subsequently called off owing to the division's inability to organize an effective bombardment in addition to intense German artillery and small-arms fire, which made the prospect of moving infantry in the open all but impossible. Unfortunately, the order to postpone the attack did not reach every unit in time, and some battalions made isolated, ineffectual attacks at heavy cost.[24] On 10 May 77e DI also lost its commander, General Barbot, who was killed by German artillery while in his command post at Berthonval Farm, thus depriving 77e DI of its driving force.

---

18 Ibid.

19 *SHD*, 22N1832; «*Compte-rendu Sommaire des Opérations de la 77e Division pendant les journées des 9, 10, & 11 Mai*».

20 *SHD*, 26N701/1; 159e RI JMO.

21 Ibid.

22 Ibid.

23 *SHD*, 22N1832; «*Compte-rendu Sommaire des Opérations de la 77e Division pendant les journées des 9, 10, & 11 Mai*».

24 26N701/1; 159e RI JMO.

The apocryphal portrayal of Barbot's death in *La Division Barbot* (told in flashback after the reader comes across the emotive '*BARBOT EST MORT!*' set in a page-break) has Barbot and 'cyclist Favier' going to observe the front from a nearby hill, despite warnings from 77e DI staff officers that the hill was being bombarded by German 105s.[25] While Favier plays the part of the ordinary soldier, hitting the dirt in the face of incoming German artillery, Barbot stands bravely ('disdainful of going to ground' as the account reads), only making evasive manoeuvres as necessary.[26] Unsurprisingly, Barbot is soon fatally wounded by German artillery fire. Atop the crest he mutters his final orders to his staff officers, and is carried off the battlefield as nearby *chasseurs* 'spontaneously' leave their trenches, line Barbot's path and present arms to honour their mortally wounded leader. Barbot turns to look at them, slowly lifting his hand to his beret in a final salute. As touching as Barbot's hero story is, it is far more likely that he was killed while in his command post – as is stated in the official reports. He was replaced by General Stirn, who had been the commander of 88e Bde (77e DI). Stirn was replaced as 88e Bde commander by Colonel Desvoyes from 159e RI. Stirn himself would fall victim to German artillery on 12 May, and would ultimately be replaced by General Pillot, who would command 77e DI for the remainder of the battle.

Despite reinforcement by 67e RI (34e DI), 77e DI found that it could not make any progress on 11 May. The division could not capture Souchez (97e RI having met serious resistance in the Souchez cemetery), and could not push on to Givenchy-en-Gohelle or Vimy Ridge without succumbing to intense enfilade fire from Souchez.[27] While the losses on 11 May were not very high, it was clear that the division would be unable to advance without a renewed and detailed artillery preparation, which it was temporarily incapable of organizing owing to the fatigue, confusion and disorder of three days constant fighting.[28] Despite meticulous planning and sound execution, which led to its substantial success, 77e DI still faced most of the same tactical obstacles that had lain before it on the morning of 9 May (Souchez, *cote* 119, Château Carleul and Vimy Ridge) at the sacrifice of their commanding officer, and the loss of surprise which had greatly facilitated their early gains. The next five weeks would be a long and difficult slog against alert German defenders well aware of the danger posed by a well-orchestrated French attack and anxious to do whatever they could to keep the French from repeating their success of 9 May.

---

[25] Humbert, *La Division Barbot*, p. 41.

[26] Ibid. «*dédaignent de se coucher*».

[27] *SHD*, 22N1832; «*Compte-rendu Sommaire des Opérations de la 77e Division pendant les journées des 9, 10, & 11 Mai*».

[28] Ibid.

34e Division d'Infanterie

34e DI (XVII CA), commanded by General Paul de Lobit, represents a radical departure from the 77e. Whereas 77e DI was an elite unit composed of excellent officers and highly motivated men, the 34e was a run-of-the-mill metropolitan division hailing from the Somme-Bionne region; on paper it was no better and no worse than any other division could expect to be. 34e DI had cut its teeth during the First Battle of Champagne from February to March 1915. The division had been situated in the Perthes region, and made a long series of costly, yet ineffective, attacks over the course of the battle.[29] Most disturbingly, many attacks after the initial push of 16–18 February were ordered as simply '*même troupes – même objectifs*' ('same troops – same objectives'). It is little wonder the Germans were uniformly prepared when 34e DI attacked during Champagne: their unchanging plan of attack left the German defenders no room for doubt.[30]

Despite the intense fighting 34e DI had endured just seven weeks before the launch of Second Artois, XVII CA was transferred from de Langle's Fourth Army in Champagne to d'Urbal's Tenth Army to take part in the Artois battle. 34e DI left Fourth Army command on 22 April and found itself, eight days later, under the command of General d'Urbal.[31] Upon arrival XVII CA received instructions from d'Urbal about the level of communication he required from XVII CA commander General J.B. Dumas. Written and verbal (telephone) reports were to be sent to Tenth Army twice a day, in addition to updated maps of XVII CA artillery placements, which were to be provided three times a month.[32] On the night of 2–3 May 34e DI began to move into its sector (in front of Roclincourt) with two groups of the AD/34 (*artillerie de division*) relieving two groups from Tenth Army reserves who had formerly held that front.[33] With the preparatory bombardment for Second Artois timed to begin on 3 May the AD/34 would have little time to acclimatize itself to the realities of a new front before taking part in the largest trench offensive the French army had yet launched.

On 2 May Dumas began issuing a series of instructions to 33e and 34e DIs. Very much like Pétain, Dumas was a hands-on corps commander who often went to incredible detail in his prescriptions. Above all else, Dumas laid a heavy emphasis on the importance of artillery–infantry liaison. He stated in one document that previous attacks had been stopped primarily owing to poor communications, and as a result the continued maintenance of communications should be a priority for both

29 *SHD*, Microfilm 9M394, 104e RI JMO.

30 *SHD*, 25N87, various documents.

31 *SHD*, 26N667/12; 88e RI JMO.

32 *SHD*, 24N741; «NOTE au *sujet des compte-rendus périodiques à fournir par les CA au Général Commt. La 10e Armée*» 30 *avril 1915*.

33 *SHD*, 24N741; «Le Général J.B. DUMAS, Commandant le 17e *Corps d'Armée au Général Commandant la 34e Division d'Infanterie*» 1 *mai 1915*.

the infantry and the artillery.[34] The artillery was instructed that its responsibility was, at all times, to know exactly where the front lines were so that they could provide quick and accurate support for the infantry. This task was to be facilitated by specially designated artillery observers whose sole responsibility would be to track the status of the front and, during an engagement (defensive or offensive), to keep the artillery informed as to the current location of French infantry.

The infantry were given a much more detailed and varied set of instructions. Dumas felt that the infantry naturally had more responsibility for maintaining liaison with the artillery owing to their mobility and the inherently subordinate role of the artillery (even in an artillery war such as the First World War). Dumas organized the distribution and use of flares, but also made sure to emphasize the continued importance of flags (which the infantry had to make themselves), and even hand and arm signals for communicating with the artillery, especially when communicating over relatively short distances.[35] Dumas laid particular emphasis on the maintenance of telephonic communication, a theme which would be repeated often in the weeks after 9 May. On 2 May he commented that the telephone network XVII CA had inherited was in adequate shape, but ordered that the number of lines be doubled if possible and laid in the walls of trenches, as opposed to along the floor, in order to better protect them.[36] Telephonic communication was to be ensured by both the frontline regiments and their attendant artillery, with the commander of the telegraph detachment responsible overall for the maintenance of XVII CA's telephone network. In this flurry of activity, however, General de Lobit scarcely made comment. He instead allowed his division to be commanded by Dumas, a situation not well disposed to providing sufficiently detailed support and direction for the upcoming battle.

General Paul de Lobit was born in 1860. A *Saint-Cyrien*, he began the war as a colonel, but was promoted to brigadier (commanding 68ᵉ Bde) in December 1914. He would rise to command 34ᵉ DI in April 1915, after the harrowing experience of 34ᵉ DI during First Champagne; Second Artois would be de Lobit's first trench battle as a divisional commander. While his performance in Second Artois was often shaky, a presumed result of his rapid promotion, General de Lobit would go on to have a reasonably distinguished career. He would command 34ᵉ DI until December 1917, at which point he was given command of the 3ᵉ *Groupe de divisions d'infanterie* in Macedonia. His actions in Macedonia would result in his being awarded the *Légion d'honneur* in December 1918. However, such heights were still far ahead of de Lobit in spring 1915. A long process of learning

---

34 *SHD*, 24N741; «Note pour les C.A.» *10ᵉ Armée, 2 mai 1915.*

35 Ibid.

36 *SHD*, 24N741; «Le Général J.B. Dumas, Commandant le 17ᵉ *Corps d'Armée aux: Généraux Commandant les 33ᵉ et 34ᵉ Div. d'Inf., Colonel Commandant l'Artillerie du 17ᵉ C.A., Colonel Commandant le Génie du 17ᵉ C.A.» 2 mai 1915* and «Le Général J.B. Dumas, Commandant le 17ᵉ *Corps d'Armée aux: Généraux Commandant les 33ᵉ et 34ᵉ Divisions d'Infanterie*» 2 *mai 1915.*

and experience would be required before the newly minted divisional commander came into his own.

On 3 May Tenth Army began its preparatory bombardment. In terms of the number of batteries available 34e DI appears, on paper, to have had an allocation of artillery roughly comparable to that of 77e DI. A deeper look, however, reveals a division struggling with outdated guns and a serious lack of munitions (especially in the days after 11 May).[37] Where 77e DI had 220mm mortars and new 37mm trench guns, 34e DI had early versions of the already outdated 58mm cannon, which were held under corps command. Where 77e DI had two batteries of modern 155CTRs under Barbot's direct command, 34e DI had to rely on 30-year-old batteries of de Bange 155L and 155Cs, which also remained under corps command.[38] To make matters worse, 34e DI, despite having an ostensibly supportive role protecting the flanks of the main thrust towards Vimy Ridge, would be attacking one of the more difficult German defensive positions, the Two Triangles. These were two triangular trench networks which created a deep field of overlapping fire into the centre of which the men of 34e DI would charge on 9 May.

Nevertheless, 34e DI would have to get by on what limited resources it had. Firing plans were ostensibly written daily by General de Lobit, although in fact Dumas played a large role in setting the agenda for each day – even to the point of designating how many batteries were to be used in the bombardment versus how many were to be retained for counter-battery fire.[39] By order of General Dumas, most of XVII CA's 23 trench guns would be concentrated in 33e DI's sector, owing to the close proximity of the French and German trenches where XVII CA's sector met with XX CA's opposite the *Labyrinthe*.[40] They would be commanded by specialist infantrymen (grenadiers) in the hopes of making their bombardment more efficient and able meet the needs of the infantry. While this would be a very welcome prospect for the men of 33e DI, 34e DI would only have access to whatever guns and shells were left over from XVII CA's arsenal.

The German defences opposite 34e DI consisted of two separate trench networks, one behind the other; both were protected by thick belts of barbed wire. As a general guideline it was estimated that each of the eight assault companies needed its own breach in the enemy wire through which to attack, and the main

---

37 *SHD*, 24N741; «*Ordre Général No. 297» XVII C.A., 5 mai 1915.*

38 *SHD*, 24N741; «Le Général J.B. Dumas, Commandant le 17e *Corps d'Armée aux: Généraux Commt. les 33e et 34e Divisions d'Infanterie, Colonel Commt. l'Artillerie du 17e C.A., Colonel Commt. le Génie du 17e Corps d'Armée. Note au sujet des canon de tranchée*» 3 *mai 1915.*

39 *SHD*, 24N741; «*Ordre Particulier No. 296 au Général Commandant la 33e Division, au Général Commandant la 34e Division, au colonel Commt.L'Artillerie*» 4 *mai 1915.*

40 *SHD*, 24N741; «Le Général J.B. Dumas, Commandant le 17e *Corps d'Armée aux: Généraux Commt. les 33e et 34e Divisions d'Infanterie, Colonel Commt. l'Artillerie du 17e C.A., Colonel Commt. le Génie du 17e Corps d'Armée. Note au sujet des canon de tranchée*» 3 *mai 1915.*

goal of the bombardment was to provide these eight 25m-wide openings, plus four additional openings in the wire of the second position.[41] The breaches were to be made primarily by the heavy artillery (*artillerie lourde*, AL) and trench guns, with some assistance from the field artillery (*artillerie de campagne*, AC), which was also responsible for counter-battery and anti-personnel fire to keep the Germans from repairing any damage done during the French bombardment. On the day of the assault the artillery would focus nearly all of its effort on the German trenches, in theory assuring their destruction. The 75s would fire at the reduced rate of four shots a minute, and would undertake a complicated series of fire and rest periods in the hopes of confusing the Germans as to the precise moment of the attack (one must wonder if this attempt at misdirection, unique to the day of the attack, served only to better warn the Germans of what was coming).[42] Ultimately, a number of 58mm and 75mm guns would be redeployed forward to provide close fire support to the advancing infantry during the attack. Goya claims that the 34ᵉ DI used a rolling barrage on 9 May which was to advance by 25m bounds, keeping a steady 250m ahead of the infantry.[43] While this seems perfectly feasible, even likely, Goya does not provide any footnotes to support the assertion, and no supporting evidence can be found in the archival sources. In any case, very little in the way of refinement was to be found in the artillery plans of the 34ᵉ DI examined.

In comparison with the artillery plan, which was relatively basic, 34ᵉ DI's infantry plan was much more detailed, and compared favourably to that of 77ᵉ DI. The infantry were to be in their jumping-off positions at 03.30 on the morning of the attack, and would be furnished with sandbags and barbed wire with which to prepare captured German positions for defence.[44] Each infantryman would carry forward with him 250 rounds of ammunition, two or three grenades (specialized grenadiers could be expected to have more) and rations for two days.[45] As with the 77ᵉ, the men of the 34ᵉ were exhorted to advance as quickly and resolutely as possible, with full faith in the ability of their neighbouring units and their reserves to cover the flanks of the assault troops. Secondary waves would follow 50m behind the first, and would include specially designated *nettoyeurs* (whose readiness de Lobit was careful to ensure in the days before the attack).[46] Engineers armed with petards and weapons very similar to Bangalore torpedoes were to go forward with the infantry to create wider gaps in the enemy wire if needed; assist in clearing enemy trenches with explosives and grenades; and coordinate the reorganization of captured trenches for effective defence against any German counter-attack.

---

41 *SHD*, 24N741; «Programme des *Tirs de l'Artillerie pendant la Période de Préparation Lente*» *5 mai 1915.*

42 *SHD*, 24N741; «*Ordre Général No. 297» XVII C.A., 5 mai 1915.*

43 Goya, *La chair et l'acier*, p. 277.

44 *SHD*, 24N741; «*Ordre Général No. 297» XVII C.A., 5 mai 1915.*

45 *SHD*, 24N741; «Instruction pour *l'Application des dispositions de l'ordre général No. 297 du Général Ct le 17ᵉ C.A.» 6 mai 1915.*

46 *SHD*, 24N741; «Note de Service» 8 *mai 1915.*

While this is admirable, it is probably to be expected of a unit preparing for battle after the distribution of Note 5779. However, 34ᵉ DI, almost entirely thanks to directives coming from General Dumas, pursued an organization well beyond the prescriptions of Note 5779. Similar to the system General Barbot designed to maintain the free circulation of men to and from the front once the battle started, 34ᵉ DI used dismounted cavalrymen to direct traffic in the rear.[47] While the use of cavalrymen, unfamiliar with the newly inherited trenches of 34ᵉ DI's sector, could not be as effective as 77ᵉ DI's using infantrymen for this task, it still represented a fair amount of foresight into the confusion that would arise on the day of the attack. Depots in the front line were well stocked with spare ammunition, grenades, flares, sandbags and all nature of materials necessary for a successful trench operation (even if this could only be done by largely depleting the reserves of these key resources formerly held in the rear of 34ᵉ DI).[48] Great attention was given to ensuring a reliable supply of food and fresh water to assault troops once combat had started, and the division was even able to organize additional training of its grenadiers in some of the newer types of grenade (including the Besozzi tear-gas grenade) only days before the battle was to be launched.[49] On paper, General de Lobit and the men of 34ᵉ DI had every reason to be confident of success on the morning of 9 May.

Only one regiment, the 88ᵉ (68ᵉ Bde), would take part in the initial assault. The other regiment, the 59ᵉ, was held in immediate reserve, while the entire 67ᵉ Bde had been placed under the command of General d'Urbal to act as army reserve, thus halving the man-power and frontage of 34ᵉ DI. The assault wave would consist of two battalions which moved into position at 03.30 on 9 May as planned. Six and a half hours later, at 10.00, the Marseillaise was played in the front trenches by the 88ᵉ RI band, and the assault battalions surged out of their *parallèles de départ*.[50] They were met with violent rifle and machine-gun fire from the Two Triangles, such that only isolated elements of the regiment could make any progress into the German trench network; the rest of the assault troops were forced to return to their jumping-off trenches (Figure 4.2). A second effort was made roughly 15 minutes after the first attack; it also met with intense small-arms fire and was very quickly repulsed.[51] Even where troops could advance across no man's land and approach the German trenches they found that most of the gaps made in the German wire had been repaired, making progress all but impossible. From 3 May the Germans had engaged in a continuing struggle to maintain the integrity of their

---

47 *SHD*, 24N741; «Instruction pour *l'Application des dispositions de l'ordre général No. 297 du Général Ct le 17ᵉ C.A.» 6 mai 1915.*

48 *SHD*, 24N741; «Le Général de Division *d'Urbal Commandant la Xᵉ Armée au Général C/t le 17ᵉ Corps d'Armée*» *5 mai 1915.*

49 *SHD*, 24N741; *«Le Général J.B. Dumas, Commandant le 17ᵉ Corps d'Armée aux: Généraux Commt. les 33ᵉ et 34ᵉ D.I., Colonel Commt. le Génie du 17ᵉ C.A.» 3 mai 1915.*

50 *SHD*, 26N667/12; 88ᵉ RI JMO.

51 Ibid.

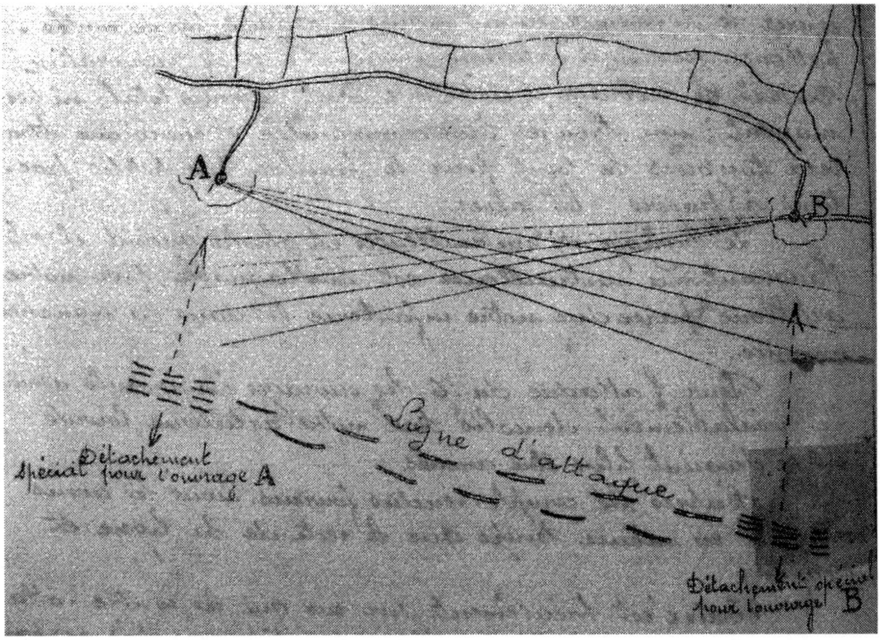

Figure 4.2 Example of German flanking fire such as 34e DI encountered at the Two Triangles

Source: *SHD*, 24N741.

position against the damaging effects of the French preparatory bombardment; on the morning of 9 May the Germans knew that they had won.

In the hope of supporting the efforts of 33e DI on his left, and after learning that elements of 88e RI had successfully captured a small section of German trenches, de Lobit organized a renewed attack to be launched at 16.00.[52] The artillery was concentrated on the Two Triangles, with the AL focusing on the left and AC on the right. The infantry attack was immediately met with a flurry of German small-arms and artillery fire, which inflicted heavy casualties on the assault battalions. When these units reached the German wire they found it intact, and were ultimately forced to retire to their jumping-off trenches under heavy German fire.[53] As with the morning attack, the failure of the artillery preparation to sufficiently damage German defences meant that the infantry had no hope of advancing. The AL had fired too far during its bombardment before the 16.00 attack, and failed to destroy the German machine guns in the left triangle. The AC, despite firing on target, was not powerful enough to inflict any serious damage on the machine guns and wire

52 *SHD*, 24N741; «*Compte-rendu Sommaire Journée du 9 Mai*» *9 mai 1915*.

53 *SHD*, 24N741; «Compte-rendu de fin de *journée*» 68*e Brigade, 9 mai 1915*.

protecting the right triangle.[54] With only one fresh reserve battalion left de Lobit declined to attack again on 9 May. 68e Bde had suffered 1,588 casualties, of which 38 were officers – including Colonel Maheas commanding 88e RI. That evening de Lobit ordered 88e RI, which had sustained roughly 50 per cent casualties (1,535 men and officers), to retire and be replaced by 59e RI, which was expected to take up the attack on the 10th.[55]

As night fell the small contingent from 88e RI that had invested a section of German trench was increasingly subjected to German heavy artillery bombardment, grenades and frequent counter-attacks, which eventually dislodged the defending *poilus*.[56] The night of 9–10 May was fairly calm, with only sporadic bursts of small-arms fire breaking the silence, which allowed a great number of wounded and others to return to the French lines during the night. The story they brought with them was discouraging. The German wire was more or less intact along the whole of 34e DI's front. Those elements of 88e RI which had managed to infiltrate the enemy trenches had done so through a German sap, hardly an opening through which any serious attack could fit.[57] The left triangle, where the men had penetrated, had not been seriously affected by the French bombardment and its machine guns were still in good order. A small number of German prisoners captured by the 88e revealed that the Germans outnumbered the 34e DI four regiments to two (a situation greatly exacerbated by the casualties suffered by 88e RI on 9 May), and had five machine-gun companies[58] in the left triangle alone.[59] 59e RI would have a difficult task before it.

On the morning of 10 May de Lobit found himself urging the colonel commanding the corps' heavy artillery to focus more of his effort on the Two Triangles, while complaining that the preparation for the 9 May attack had been wholly inadequate.[60] This process, combined with the antiquity of the heavy guns available to XVII CA, demonstrates the enormity of the gap between 77e DI and 34e DI in terms of artillery support. Barbot and his successors could order their organic heavy artillery batteries to fire on whatever targets they chose; de Lobit had to submit requests for artillery support, which would be weighed against similar and competing requests from 33e DI, and even from the neighbouring X and XX CAs. At 14.30 a general note to 33e and 34e DIs was issued by the colonel commanding XVII CA's artillery, who blandly stated that the artillery preparation had been completed and had conformed to the prescriptions set out for the bombardment on

[54] *SHD*, 24N741; «*Compte-rendu Sommaire Journée du 9 Mai» 9 mai 1915.*

[55] Ibid.

[56] *SHD*, 24N741; *«Compte-rendu des* événements *de la nuit»* 34e *DI, 10 mai 1915, 04.00.*

[57] Ibid.

[58] Twenty machine guns, four per company.

[59] *SHD*, 24N714; «Le Lieutenant-Colonel Schneider *Commt le 88e Régt. D'Infanterie à M. le Colonel Commt la 68e Brigade» 10 mai 1915.*

[60] *SHD*, 24N714; «Note» 34e *DI, 10 mai 1915.*

9 May.[61] The attack was to be renewed that afternoon and would be preceded by a five-minute bombardment by each division's respective field artillery.

Meanwhile, the commander of the AD/34 was seeking to forge a closer liaison between the infantry and the heavy artillery. A meeting between representatives of the AL, AC and the infantry was planned for 06.00 on 11 May in order to pinpoint key problem areas, such as machine-gun nests. Together the infantry and artillery would determine how best to handle each target; how many guns to use and of what calibre; and how many shells to devote to the systematic destruction of individual components of the German defensive network.[62] Maps would be drawn up based on the results of the meeting, and copies would be passed on to de Lobit, who would offer input on the fire plan and alter it as necessary. As laudable as this effort was, it was nevertheless a week too late, and 59ᵉ RI still had to attack the Two Triangles on the 10th.

At 04.30 on 10 May 59ᵉ RI took over the front formerly held by the battered 88ᵉ. Throughout the day 59ᵉ RI had been subjected to heavy bombardment by German 105s and 150s. When the 59ᵉ finally did attack at 17.00 it was met with an intensified counter-barrage, in addition to thick small-arms fire.[63] The assault battalions, whose morale had been sapped upon hearing of the casualties suffered by 88ᵉ RI the day before, quickly returned to their trenches after suffering only 48 casualties.[64] A second attempt was made at 18.00 with the same results, a rapid and forceful German response which drove the French infantry back into their *parallèles de départ* post-haste. As even the slightest movement along the front was met with a flurry of German fire, no further attempts were made to launch an attack that day. The frustratingly poor performance of the entire corps had not escaped the attention of d'Urbal, who insisted that the soldiers of XVII CA were failing not due to material constraints, but merely out of a lack of élan.[65]

It was clear in light of the previous two days that XVII CA's methodology was simply not working. For the attack on 11 May, 34ᵉ DI was given access to all of the 155mm batteries XVII CA had available, including a battery of 155CTRs. The infantry was instructed to enter no man's land during the preparatory bombardment, thus reducing the amount of ground it would have to cover once the bombardment finished.[66] There is no evidence of any sort of rolling barrage being provided. 34ᵉ DI's plans were nearly derailed at 10.00 when de Lobit learned that 19ᵉ DI (X CA) on his right would not be making a full attack and, furthermore, would not be linking up with his own efforts, thus leaving his right flank exposed

---

61 *SHD*, 24N714; «*Ordre Particulier aux Généraux Cdt les 33ᵉ et 34ᵉ Divisions, au Colonel Cdt l'artillerie*» 10 *mai 1915, 14.30.*

62 *SHD*, 24N714; «Note» AD 34, 10 *mai 1915.*

63 *SHD*, 24N714; «Compte-rendu de fin de *journée*» 68ᵉ *Brigade, 10 mai 1915.*

64 Ibid.

65 *SHD*, 24N714; «Note *du C.A.*» *11 mai 1915.*

66 *SHD*, 24N714; «Instruction *pour l'application des prescriptions de l'ordre No. 302*».

to German enfilade fire.[67] Two and a half hours later, after discussing the situation with Dumas, de Lobit wrote to the general commanding 19e DI to accept that the two would be making separate attacks, but asked that 19e DI artillery fire on the right triangle to assist the attack of 34e DI.[68] De Lobit pointed out a series of machine guns in the right triangle which were especially problematic and which had caused grievous harm to 88e RI by firing into its exposed right flank on 9 May.

The assistance of 19e DI would be crucial not only as a means of protecting the right flank of 34e DI, but also by providing extra weight to the preparatory bombardment. Unfortunately, the extra guns provided by 19e DI made little impact. The attack was launched by 59e RI at 14.00, and failed miserably – as had the attacks of the previous two days. The men of the 59e managed to advance less than 50m before intense German fire forced them to go to ground and eventually retire.[69] As French reserves moved forward they marched into an intense and on-going German counter-barrage which ultimately halted a second attack made at 15.25.[70] These tentative attacks had exacted a heavy toll on the men of 59e RI and were being supported by an ever-decreasing quality of artillery preparation owing to a lack of munitions: by 16.00 on 11 May 34e DI was reduced to only 60 shells per gun.[71] De Lobit found himself desperately begging Dumas for resupply, which failed to materialize. He even begged the neighbouring 33e and 19e DIs for ammunition on loan, which de Lobit promised to repay once he had been resupplied; this failed to provide 34e DI with any more shells. Nevertheless, a third attack was planned and ultimately launched at 18.00. This attempt was met with a strong German counter-bombardment from 17.50 which caused the attack to sputter and fail.[72] Some units advanced roughly 100m into no man's land and began to dig a new trench there with the assistance of a small group of engineers, the only concrete achievement made that day.

Comparison

On the ground 34e DI cannot claim to have made any real progress. An advance into no man's land does not require a full-blown attack; nor is it particularly impressive, and at no time were the Two Triangles under any serious threat of being overrun. There is no doubt that 77e DI's experienced and competent leadership, access to organically attached modern heavy artillery, ample ammunition and the support of two expert divisions on either side of it (70e DI and the DM) gave the division

67 *SHD*, 24N714; «Compte-rendu» 34*e DI, 11 mai 1915, 10.10.*

68 *SHD*, 24N714; «Le *Général de Lobit Cdt la 34e DI à M. le Général Cdt la 19e DI» 11 mai 1915, 12.45.*

69 *SHD*, 24N714; «Compte-rendu de fin de *journée*» 68*e Brigade d'Inf., 11 mai 1915.*

70 Ibid.

71 *SHD*, 24N714; «*Compte-rendu et renseignements*» 34*e DI, 11 mai 1915.*

72 *SHD*, 24N714; «Compte-rendu» 34*e DI, 11 mai 1915, 21.00.*

a very good chance of success on 9 May. The questions raised then are: could 34e DI have met with similar results if it had had access to the same level of support. Would access to more shells and better guns have led 34e DI to develop a more complicated and effective preparatory bombardment and subsequent barrage? Could better leadership have produced correspondingly better results for 34e DI in the opening phase of Second Artois, or was it too much to ask a unit to launch an attack barely more than a week after taking over its new sector?

At first glance it would seem reasonable to suggest that if 34e DI had had access to more modern, quick-firing heavy artillery and a greater supply of shells it would have performed demonstrably better on 9 May and the subsequent days. One of the most crucial problems that 34e DI consistently ran into was an inability to destroy the German barbed wire. Indeed, when small breaches were made they were quickly repaired by the Germans, who could work with relative impunity owing to the French lack of munitions, which made it very difficult for their artillery to deter German soldiers from working. Better heavy guns would have made the gaps in the enemy wire bigger, which would make them harder to repair, while an increase in the number of 75mm shells available would make it easier for the French to impair any German reconstruction efforts by maintaining a steady bombardment of the enemy trenches. A greater supply of ammunition might also have encouraged 34e DI to plan an effective rolling barrage[73] and to fire longer, more effective bombardments immediately preceding infantry attacks (some bombardments were only five minutes long). This stronger, more prolonged bombardment would likely have had a greater effect on German morale that, combined with the above, would have made success much more likely.

Even in an ideal situation an army can never have an infinite amount of supplies. Ultimately, some sectors will be prioritized over others and receive more material weight, and better troops, in relation to their increased importance to the operation or, indeed, to the war effort as a whole. While the situation of 34e DI is symptomatic of an army still in the grip of a general shell shortage – and was coordinating the introduction of new arms (especially modern heavy artillery) while phasing out obsolete models – it was still, first and foremost, a victim of being assigned an auxiliary role. The main task of 34e DI was not to capture the Two Triangles. Rather, the main task of 34e DI was to cover the flank of the attack on Neuville, and in that task one can argue that 34e DI was at least marginally successful. It is debatable whether the capture of the Two Triangles on 9 May would have had any impact on operations along the main axis of attack whatsoever.

The events on 34e DI's front might be called a 'wanton waste of life', the result of poor leadership on the part of callous generals, and will no doubt be seen as such by many who read of their experiences. The under-equipped and unsuccessful 34e DI poses a great problem to historians, one which falls outside the domain of moral judgement. Namely, how does the experience of such a division

---

[73] One might argue that 34e DI was not aware of how to fire a rolling barrage, but the technique was certainly in use by XVII CA units during First Champagne: *SHD*, 25N87.

fit into the on-going arguments about the improvement of Allied skill whereby French, British and later American troops launched increasingly successful attacks against the Germans as the war progressed? This evolution is talked about as if entire armies were improving as a whole by trial, error, reflection and foresight. However, how does one measure this progression? Does one look at a skilled unit practising the military art at a high level (like 77ᵉ DI), or does one look at a less successful unit like 34ᵉ DI? Even then, should victory on the field of battle determine how 'skilled' a unit is considered to be, or are there other criteria such as the efficacy of leadership, the efficiency of the bureaucratic machinery within a unit, access to materiel and logistical networks, or the élan of its soldiers? Millett and Murray would argue that victory is not the sole determiner, but one still must take actual results into account.[74]

As has been argued in this chapter, 34ᵉ DI was not as far behind 77ᵉ DI in terms of leadership and organization as one might expect. It was not quite as efficient, and it had its own bureaucratic quirks which made certain aspects of command and control less efficient (especially the maintenance of the heavy artillery under corps command); but its planning and preparation still seem to have warranted better results than the division actually produced from 9 to 11 May. If one does choose to focus on the intellectual and thus, implicitly, the doctrinal state of a fighting force, how and when is the quantitative result examined? Can a unit be theoretically sophisticated but sufficiently poor in execution to be defeated? At what point does a commander's thought pass from being forward-thinking to naïvely optimistic or fanciful? When and how does one measure the strength of the enemy *vis-à-vis* the units one wishes to study; and how does decide to what extent enemy activity distorts the efficacy or inefficacy of a given unit? There can be no definitive answers to these questions. Success on the field of battle must, in the end, be the criteria for greatness, but the intellectual preparation which made success possible (even if victory was not always achieved) must remain an important part of the equation. When the 77ᵉ and 34ᵉ DIs are revisited in Chapter 7 we will explore more fully the intellectual aspects of war: the preparation and reconstruction which occurs after 11 May and which constitutes the foundation from which the 16–18 June attack would eventually be launched. After this examination, many of the questions raised here can be addressed, and some conclusions about infantry combat in 1915 can be reached.

[74] Millett and Murray, *Military Effectiveness*, pp. 2–4.

Chapter 5
Grignotage (12 May–12 June)

It is easy to conceive of operations in the First World War as merely being made up of their more spectacular moments, the Battle of the Somme's opening day being the most obvious example. Conceptualizing long and steady operations, which took place over the course of months, is not only very difficult, it can also be exceedingly tedious: both reasons that many histories gloss over these periods as either inactive or, at worst, frivolous and wasteful (especially if many lives are lost doing 'nothing'). Such interpretation does a great disservice to the complexity of operations in the Great War, and neglects the 'real' war: the war of logistics, of engineering and of attrition. Shells needed to be produced, counted, delivered and then mathematically applied in complex bombardments of enemy defences. Bombardments and counter-bombardments destroyed trenches which needed repair, and the threat of bombardment encouraged the ever more complex organization of trench networks to protect the men deployed to defend them. It was a seemingly indefatigable machine of manpower, horsepower and high explosives that was the engine of the Great War, an engine that would only occasionally and inconsistently belch out a 1 July 1916 or an 8 August 1918.

To ignore the great periods of what is considered 'inactivity' is also to gloss over the reflection and evolution which happens after a big offensive which frequently yielded results as important as the battle which it followed. While the initial genesis of new ideas and methodologies most frequently comes from the bottom up, these ideas are utterly inert if the bureaucratic structure of the army does not pick them up and diffuse them to other, less innovative units. Once ideas were diffused there was still left the task of training, absorption and practice. New soldiers brought in to replace losses would effectively deskill shattered units. Recruits needed to be acclimatized to trench warfare, and given practical training in how things actually work within a given unit before they could be expected to take part in any substantial engagement. This, of course, is not even to mention the innumerable small, local offensives and trench raids which occurred during quiet periods as units jockeyed for a better position *vis-à-vis* the enemy in terms of territory and intelligence.

After 11 May Tenth Army did not lie dormant, but continued to flex its manpower and brainpower with the goal of reconstituting its strength and preparing itself for a second attempt to take Vimy Ridge. Captured positions needed to be consolidated and linked up with the old French front lines; battle-weary units needed to be relieved, rested and reconstituted with fresh recruits. Further back, large elements of the logistical apparatus (hospitals, depots, railway lines, and headquarters) all needed to be moved closer to the new front line, particularly in

XXXIII *Corps d'Armée* (CA's) sector. New artillery positions needed to be built and the guns stationed there needed to re-register on a range of enemy targets, some new and some old. While 77e and 34e *Divisions d'Infanterie* (DIs) found themselves in radically different situations on 12 May (77e DI inflated by reserves, but weary from casualties and still reeling from the loss of Barbot; the 34e dejected and demoralized from their abject failure during the previous three days) they both had to deal with similar situations, the study of which provides some insight into the experience and workings of a French infantry division in the First World War.

The Second Phase – Fighting for a *Base de Départ*

The second phase of Second Artois was a long, slow struggle fought mainly by XX CA, XXI CA and 70e DI to secure the flanks of the 77e and Moroccan divisions (DM) from the deadly enfilade fire pouring out from Notre Dame de Lorette (NDL), Souchez and Neuville, which had so effectively stopped the 11 May attack. 12 May, and subsequent days, would be largely spent transforming the captured German trenches and *boyaux* into an integrated French defensive network and relieving battered units. 97e *régiment d'infanterie*, RI (77e DI) would be replaced by 14e RI starting on 13 May as the 77e began organizing its sector in such a way as to maximize flanking fields of fire for its machine guns and provide extra storage space for food and water, the lack of which had been sorely felt over the past few days.[1] The relief of the DM would continue. 78e Brigade (39e DI) would spend the 12th trying to recover from the disaster of the previous day. Despite heavy losses and having poorly organized positions, 77e, 18e and 39e DIs were all expected to engage in certain small operations while relief and rebuilding continued.

Along the left flank of XXXIII CA progress continued to be made. On 12 May 70e DI would again attack *bois* 125 and Carency in liaison with 61e *bataillon de chasseurs à pied* (BCP) from 77e DI, this time with greater success despite 61e BCP having sustained heavy losses repelling a German counter-attack the night before. Advancing infantry units were well provided with engineers to help shore up captured trenches, and ample flags and flares were provided with which to liaise with divisional artillery, which included two batteries of 155CTRs.[2] Moving rapidly, and supported by a much longer and more precise artillery bombardment (slow cadence preparation started at 11.00 and lasted until 16.00 when the artillery fired an intense hurricane bombardment for an hour immediately preceding the attack), 70e DI captured both *bois* 125 and Carency. Owing partially to the strong support of 70e DI, as well as a large reinforcement of four battalions, 13e DI also made good progress on 12 May. The division would go on to capture the chapel of

[1] *SHD*, 22N1808; «*77e Division à 88e Brigade, 67e Brigade, Chasseurs, 97e R.I.*» *13 mai 1915, 16.30.*

[2] SHD, 24N1754; «*Ordre pour la Journée du 12 Mai 1915*» *70e DI, 07.00.*

Notre Dame de Lorette during a night attack and clearing the Germans from the southern half of that important hill.[3]

The success of 13e and 70e DIs had placed the Germans holding Ablain-Saint-Nazaire in a deep salient in which they were subjected to French small-arms fire from the front, both flanks and rear. Responding to this situation, the defending Germans pulled out of the village to a defensive position resting on the Château Carleul, Souchez, the Ablain cemetery and the *sucrerie*.[4] German troops on NDL also retired in order to provide a strong, contiguous line of defence with the units retiring from Ablain. 70e DI would begin cautiously occupying Ablain the next day, while 77e and 13e DIs made a pincer attack on the *sucrerie*. Engineers were busy repairing communications and expanding the trench network of the newly captured Carency, *bois* 125 and Ablain with a view to placing 70e DI in a good position to take part in the attack on Souchez on the 14th. Owing to the poor progress of XXI CA, which suffered flanking fire from freshly dug German defences north of NDL on 13 May, Pétain would be forced to postpone his planned assault on Souchez.[5]

Despite the heavy losses suffered by XX CA on 11 May, 11e DI, and elements of 39e DI, continued their attack on Neuville the following day. The attack was supported with a four-hour bombardment which would prove insufficient to render the German positions in Neuville vulnerable to infantry attack. 319e RI, the northernmost unit from 39e DI that would attack on 12 May, lost fully 50 per cent of its manpower, mostly to machine-gun fire, during its abortive assault.[6] To the right of 319e RI the rest of 77e Bde made slow, yet steady progress with a heavy expenditure of grenades and trench mortar bombs, only to be forced back owing to the dismal performance of 319e RI, which had failed to cover the left flank of the attack effectively. Likewise, 11e DI was stopped by German defences in the *Labyrinthe* and Neuville. General Nourrisson (commanding 39e DI), horrified at the losses his unit had suffered, pleaded in a letter to XX CA headquarters that no further general attacks be ordered, and that only small, partial offensives on clearly defined objectives in Neuville continue until his division could effectively reorganize itself. His concerns would not be heeded by d'Urbal, who demanded that offensive action continue at the same time as new trenches and *boyaux* were being dug, old ones repaired, and fresh infantry units were fed into the maelstrom.

Medium-scale, rushed attacks continued with similar patterns of minor successes contrasting with costly failures until May 15. On this day a more general,

---

3 SHD, 22N1808; «*Ordre Général d'Operations No. 456 pour la journée du 12 Mai (1ère partie)» 11 mai 1915, 23.55.*

4 SHD, 24N1754; «*Ordre d'Operations pour la Journée du 13 Mai» 70e DI, 13 mai 1915, 10.00.*

5 *AFGG, Tome III*, p. 53.

6 SHD, 22N1357; «*Compte-rendu complémentaire des* événements *de la Journée du 12 et de la nuit du 12 au 13» 39e DI; 13 mai 1915.*

if uncoordinated, attack across the whole front met uniformly with disaster.[7] Reserves had been pushed as far forward as possible in the hopes that a 9 May-style breakout could be turned into a sustained breakthrough. Unfortunately, Tenth Army had been steadily losing guns since 9 May owing partially to faulty ammunition, which led to barrel explosions. This greatly reduced the density of the artillery preparation. The bombardment was further compromised by Tenth Army's slavish adherence to a largely unchanging artillery plan with which the Germans were well acquainted by 15 May.[8] The situation was made even worse by a heavy influx of German reinforcements in the area of Souchez, which included the addition of a series of new machine-gun emplacements installed on 13 and 14 May.[9] After the two-hour artillery bombardment troops across the front of Tenth Army rose out of their trenches, and were met immediately with a thick bombardment from German 77s and 105s as well as intense German rifle- and machine-gun fire. The units of Tenth Army were severely undermanned, outgunned, tired and disorganized; they were simply not capable of such an ambitious undertaking against fresh German troops after nearly a week of uninterrupted combat.

Appalled by the continued inefficacy of d'Urbal's methods, Foch demanded an end to the hurried offensives, which had been Tenth Army's operational method since 11 May.[10] Foch stated that Tenth Army 'can no longer make hurried attacks, but [should pursue] a complete and detailed preparation similar to that which preceded the attack of 9 May. Such a preparation demands time.'[11] Foch estimated that it would take at least eight days to reconstitute and resupply Tenth Army to the point where another general attack could be launched. Until that time, the Tenth Army should pursue partial offensives launched with massive artillery support on clearly designated objectives. Perhaps encouraged by the success of 70ᵉ and 13ᵉ DIs in capturing Ablain-Saint-Nazaire, Foch urged d'Urbal to attack targets 'by actions from the front and the flanks, at suitable distances with [the support of] heavy, field and trench artillery. All precautions will be taken to render such preparations effective.'[12]

Between 15 May and 15 June Tenth Army launched a never-ending series of small, partial offensives whose goal was to eliminate the problem areas on

---

7 AFGG, Tome III, p. 56.

8 SHD, 22N1808; «Note pour les *Généraux Commandant les C.A. et l'Artillerie»* 17 *mai 1915, 09.15.*

9 SHD, 24N1754; «Lt.-Col. *Bonet à Col. Grange» 14 mai 1915, 08.30.*

10 Daille, *Histoire de la Guerre Mondiale*, p. 118 and Doughty, *Pyrrhic Victory*, p. 162.

11 *AFGG, Tome III*, p. 58: «ne *peut plus* être *le fait d'une offensive brusquée, mais bien d'une préparation complète et minutieuse, analogue à celle qui à précédé l'attaque du 9 mai. Cette préparation demande du temps.»*

12 Ibid., p. 59: «par des actions de front et de *flanc, à bonnes distances, d'artillerie lourde et de campagne ou de tranchée. Toutes précautions seront prises pour rendre cette préparation efficace.»*

the flanks of the 77e and Moroccan divisions. Attacks were repeated for days or in some cases weeks on end in which the same troops were asked to attack the same position, using the same methodology, time and again. On 23 May XXI CA succeeded in capturing the entirety of Notre Dame de Lorette. Four days later, on 27 May, 70e DI captured the Ablain cemetery and then proceeded to take the *sucrerie* on the 31st. These successes would ultimately place 70e DI in a good position from which it could attack Souchez from the west, thus greatly improving the prospects of 77e DI which was attacking Souchez from the south.

By vastly shortening the fronts over which attacks were being launched the French addressed a major problem which had dogged their 9–11 May attacks: a lack of artillery and munitions. Attacks on narrow fronts could receive a much more dense and detailed artillery preparation than could general offensives in which heavy artillery (*artillerie lourde*, AL) batteries might be asked to fire on entirely new sets of targets two or three days in a row. The sole attempt at a larger-scale attack in this period (on 25–26 May) met uniformly with failure as IX, XXI and XXXIII CAs found themselves attacking without the weeks of pre-planning, the fresh troops and the benefit of surprise which had made 9 May such a success.[13] While the launching of small, local attacks as opposed to general offensives largely addressed the material problem facing the French in Artois, it did little to address the lack of expertise exhibited by too many units.

From mid-May until early June there was a long dialogue between General Maistre and his subordinate commanders in which Maistre did everything in his power to try to get his corps to use more effective methodology in all aspects of trench warfare. This is most readily apparent in his constant urging of the soldiers of XXI CA to vary their artillery bombardments, even if just a little, in the hopes of achieving surprise against the German defenders. Maistre had to repeat this recommendation three weeks in a row, despite similar recommendations distributed army-wide by d'Urbal which were to the same effect. Maistre also had to issue a variety of directives to his men exhorting them to not make easy targets of themselves 'by forgetting the most elementary precautions against aerial observation and not executing instructions which had often been repeated on the subject'.[14] Resupply missions were being undertaken in broad daylight; men remained in trenches less than 30m from the enemy, which led to many instances of losses from friendly fire during artillery preparation; troops and horses would frequently congregate in open areas and casually traverse clearly visible roads. Ultimately, Maistre would have to indicate exactly how many men or horses could be present at any one time along a whole list of various crossroads and rear areas along the XXI CA front.[15]

---

13 Doughty, *Pyrrhic Victory*, p. 162.

14 SHD, 22N1472; «Le Général Maistre Commandant le 21e *Corps d'Armée à M. toutes les autorités»* 30 *mai 1915*: «par *l'oubli des précautions les plus* élémentaires *contre l'observation aérienne, et l'inexécution des prescriptions souvent répétés à ce sujet»*.

15 SHD, 22N1472; «Instruction *Générale No. 46» 21 mai 1915*.

Offensively, the French had a number of issues to deal with as well. Grenades had proved vital in attacking heavily defended urban areas, yet a large proportion of French infantrymen were not trained to use the weapon effectively. Maistre noted a serious deficiency in his men's use of grenades, which was made all the more frustrating by the fact that several XXI CA attacks had been checked by German superiority in this field.[16] In an attempt to rectify the situation he increased the number of grenades per division, mandated that all infantrymen (as opposed to just grenadiers) should carry and be trained in the use of grenades, and also circulated documentation on how to improvise grenades from common trench materials.[17]

Despite Foch's putting an end to the hasty general offensives that d'Urbal had been ordering between 10 and 15 May those Tenth Army units that were still engaging in offensive action were doing so at a tempo which simply outstripped their logistical and tactical abilities. 70ᵉ DI, after having lost as many men from 9 to 13 May as the entire XX CA lost on 9 May (roughly 2,500 men, including seven battalion commanders), was in a state of perpetual action for the entirety of Second Artois.[18] Fayolle, despite having access to generous amounts of modern, quick-firing heavy guns and elite *chasseur* battalions, still felt pushed too hard, regularly noting sentiments in his diary such as: 'Night-time, order is to attack the trench network east of Ablain … We are not ready, nothing's been done, yet the order is firm. Foch demands it … He doesn't listen anyway and follows his own ideas.'[19] Despite these pressures, and being faced with one of the toughest defensive networks on the entire front, 70ᵉ DI continued to advance at a steady pace and was one of the most successful divisions in all of Second Artois. Other units, including XXI and XX CAs, made much slower progress against similarly difficult defensive networks. XVII and X CAs, on the other hand, performed so poorly that on 18 May d'Urbal actually called for both corps to be replaced in their entirety by 24 May, an order he quickly rescinded.[20]

Faced with myriad difficulties, the French command structure did not sit idly by. On 18 May Joffre sent a letter to Tenth Army confirming what Foch had argued three days earlier when he intervened in d'Urbal's operations: Vimy Ridge would not be taken in any improvised attack hoping to recapture the magic of 9 May.[21] Joffre stated his opinion that the principal army corps involved in Artois were

---

16 SHD, 22N1472; «Note de Service» 20 *mai 1915* and «Instruction *indiquant les exercices à suivre pour l'instruction des grenadiers» 30 mai 1915.*

17 SHD, 22N1472, various documents.

18 Fayolle, *Cahiers secrets de la Grande Guerre*, p. 103.

19 Ibid., p. 106: «Le *soir, ordre d'attaque sur l'îlot est Ablain ... nous ne sommes pas prêts, rien n'y fait, l'ordre est ferme. C'est Foch qui l'exige ... Il n'écoute d'ailleurs pas et suit son idée.»*

20 Palat, *La grande guerre sur le front occidental*, p. 303.

21 SHD, 16N1677; «Note sur le situation le 18 *mai 1915 des armées alliées dans le nord de France» no. 3017.*

all exhausted and needed time to rest and refit.[22] Thus, the battle was recognized even at the highest levels as having entered its second phase of small, piecemeal attacks against locally important points of terrain from which would eventually be launched another general offensive. The same transformation had occurred in Champagne a few months earlier and would come to be the normal trajectory for most large-scale, extended trench offensives for the rest of the war.

Two days later, on 20 May, *Grand Quartier Général* (GQG) published its first reflections on the fighting in Artois. The document stated that the fighting in Artois had 'entirely confirmed the value of the principles and procedures laid out in the note of … 16 April 1915, notably in regards to the importance of a detailed preparation'.[23] The note comes out firmly in favour of the 'continuous battle' idea purported by General Castelnau and others, as opposed to the 'methodical battle' favoured by Foch, stating that 'the essential lesson is the *necessity* for *pushing the reserves as far forward as possible* … the engagement ought to be organized to advance in one continuous and automatic push'.[24] This would be the dominant methodology until after the 25 September offensives, which would prove the utter inefficacy of a continuous approach. The concept of very large and mutually supportive battles was raised yet again, with the document stating that such battles would force the Germans to disperse their forces; no allusion to the fact that the French must disperse their own forces to launch such attacks is present.[25] That said, the document does allow for the small, partial offensives which had been the norm in Artois since 11 May, provided such attacks are launched to gain important pieces of terrain from which to eventually launch a general offensive.

The most prescient statements made in the document concern the defence in trench warfare. What might be called the 'German way of defence' is well described, and gives a good indication of just how formidable the field defences were becoming on the Western Front:

---

22 Ibid.

23 SHD, 16N1677; «Premier *enseignements à tirer des combats récents*» 20 *mai 1915; no. 3019*: «Les *résultats des combats engagés depuis une semaine au Nord d'ARRAS ont entièrement confirmé la valeur des principes et des procédés exposés dans le notes sur les attaques du 16 Avril 1915, notamment en ce qui concerne l'importance d'une préparation minutieuse.*»

24 Édouard Castelnau was one of the more influential, but lesser-known, French generals of the Great War. Commander of Second Army at the time of Second Artois, he would go on to be the first commander of the *Groupe des Armées du Centre* (created in June 1915). From there he would be appointed Joffre's chief of staff from late 1915 through 1916. *SHD*, 16N1677; «Premier *enseignements à tirer des combats récents*» 20 *mai 1915; no. 3019*: «*L'enseignement essentiel est la nécessité de pousser les réserves en avant le plus possible … l'engagement doit* être *organisée d'avance en vue d'une poussée continue et automatique.*»

25 SHD, 16N1677; «Premier *enseignements à tirer des combats récents*» 20 *mai 1915; no. 3019.*

> The first line of German trenches is only occupied by a small number of men and their defence is assured by a large number of machine guns … specifically deployed to realize flanking fire. The infantry has a number of deep dug-outs. Thus, the destruction of casemates and dug-outs ought to be the principle objective of the bombardment by artillery and heavy mortars.[26]

Realizing the strength of the German system, which had been developing in the active sectors along the Western Front (and was to be fairly standard by 1916), GQG insisted that French positions be organized so that sections of front were defended not by a trench full of riflemen, but by a series of mutually supportive centres of defence heavily armed with machine guns.[27] While *poilus* on the ground would largely resist implementing a defence in depth, the idea was well known from 1915 onwards, offering yet another example of late-war practice being conceived of and practicable as early as spring 1915.

Two days after releasing this follow-up to Note 5779, GQG received a letter in response from the *Groupe Provisoire du Nord* (GPN) stating that they could not possibly move reserves forward as fast as GQG hoped without extensive motorized transport.[28] In response, GQG authorized the transfer of 18 groups of automobiles to the GPN.[29] These 18 groups were split into three equal *groupements*, each of which could transport a brigade of infantry. While on paper this was a positive step towards increased motorization, it was ineffectual in practice. The automobiles could not transport reserves through the old trench networks, which meant reserves would still have to walk several kilometres before reaching the front lines. This was recognized by Joffre, who stated that he envisaged these automobile groups being used to transport troops laterally, which of course did nothing to assist the GPN in getting troops forward more quickly. It would take another year before the automobile would come into its own as an integral piece of the logistical chain during Verdun and the Somme.

Within Tenth Army a range of procedures were updated based on the experiences of Second Artois in a more personalized effort to refine the French tactical methods in everything from the need for proper personal hygiene to the appropriate laying of telephone wires, and units were given training in a wide range of trench skills that had clearly been lacking. A massive influx of men and materiel flowed into Tenth Army, and IX CA was shifted into a position between XXXIII and XX CAs in an effort to give more punch to the second attempt at a

---

26 Ibid. «Les *tranchées de premier ligne allemandes ne sont occupées que par un petit nombre des hommes et leur défense est principalement assurée par une grande quantité des mitrailleuses … particulièrement disposées pour réaliser des feux flanquants. L'infanterie dispose de nombreux abris profondément enterrées. La destruction de ces casemates et abris doit donc* être *l'objectif principale du bombardement par l'artillerie et les gros lance-bombes.»*

27 Ibid.

28 SHD, 16N1677; «Note pour le Commandant *en Chef» 22 mai 1915; no. 3030.*

29 Ibid. «Note pour M. le Général Foch Commandant le G.P.N.*» 23 mai 1915; no. 3034.*

general offensive on 16 June. Despite heavy loss and a slow advance over the course of four weeks Tenth Army had secured many of the problem areas on the flanks of the main thrust towards Vimy Ridge (with the glaring exceptions of both Souchez and Neuville), and seemed poised finally to take that dominating piece of terrain.

Partial Offensives: 77ᵉ DI after 11 May

The period from 11 May to 15 June would be very trying for 77ᵉ DI, and serves to underscore the difficulty which even elite units had in assaulting urban positions – especially while suffering from an imperfect tactical and logistical organization. After having lost General Barbot on 10 May the division lost his replacement, General Stirn, two days later on 12 May.[30] A German high-explosive shell had struck his command post at Berthonval Farm only hours after he had been given the rank of general commanding 77ᵉ DI, thus depriving 77ᵉ DI of an able and intelligent commander for the second time in as many days. An hour later General Delmotte of 67ᵉ Bde (34ᵉ DI), which had been sent forward from Tenth Army reserves to reinforce the 77ᵉ, took command of the division and ordered 83ᵉ RI (67ᵉ Bde) to relieve the *chasseurs* who had made a failed attack earlier that day and were exhausted from four days of consecutive action.[31] Command of the division was officially passed over to General Pillot the next day, with Delmotte later taking over tactical command of 77ᵉ DI sector.

From 10 May German artillery maintained an almost ceaseless bombardment of the 77ᵉ DI position, which greatly slowed the consolidation of their trench network. Nevertheless, the division continued to make small local attacks launched by the relatively fresh men of 67ᵉ Bde.[32] These attacks, ostensibly made by regiments but in reality never amounting to more than a battalion or so, invariably saw French infantry capture a section of German trench, only to be pushed out by subsequent German bombardment and counter-attack. This continued until 16 May when the division was ordered to pass command of 77ᵉ DI's BCPs to Fayolle's 70ᵉ DI (although their provisions were still to come from 77ᵉ DI depots) and to use 77ᵉ DI artillery to support 70ᵉ DI attacks.[33] Thus *La Division Barbot* entered the role it would fill for the majority of Second Artois, that of a defensive unit largely concerned with improving its trenches and *boyaux* while its artillery supported the actions of 70ᵉ DI as it and the 13ᵉ ensured the capture of Notre Dame de Lorette, and began to approach Souchez from the west and north-west.

The *Journal des Marches et Opérations* (JMO) for 77ᵉ DI sums up the period from 18 May to 15 June in only a few short pages. The story told is one of constant

30 SHD, 26N407/2; 77ᵉ DI JMO.

31 Ibid.

32 Ibid.

33 Ibid.

bombardment by the Germans and relatively heavy losses. In the first week an average of 50 men a day were lost while the division dug two new *boyaux* which would connect the front and rear trenches.[34] As can be expected, Pétain lost little time in getting his divisions working hard again to prepare for any future action. On 17 May he ordered 18ᵉ and 77ᵉ DIs (neither of which were making any attacks) to refine and reorganize their artillery methods.[35] Observation zones were to be expanded and improved wherever possible; registration on all known enemy positions was to be completed, documented and distributed in order to make any artillery action more efficient; and better liaison with the infantry (especially telephone lines) was to be sought so that the artillery could provide immediate and effective support in case of a German attack.[36] In an effort to help the artillery engage in counter-battery fire more efficiently Pétain (via his chief of staff) also asked that the infantry be more proactive in reporting any German artillery action.[37] These reports were to include the date and time of observation, the target of enemy fire, the type of bombardment or barrage, a rough estimate of the number of shells fired and the believed location of the enemy battery, or batteries, involved. The note stressed that the heavy artillery could not effectively support the infantry without the infantry doing everything it can to help.[38] This was undoubtedly a response to the poor terrestrial observation available to XXXIII CA. Many key German positions, including *cote* 119, could not be seen from XXXIII CA positions, thus forcing the corps to rely almost entirely on aerial reconnaissance.[39]

Pillot added to this programme of reorganization by insisting that artillery batteries place themselves in less-exposed positions as casualties, especially to horses, were mounting; this order would be expanded to include resupply being restricted to night-time by Pétain, who also expressed concern about the mounting horse casualties.[40] While the loss of horses posed a growing logistical problem, it also posed a health problem. Many horses were being left where they died and not buried, which threatened to spread disease in the crowded trenches. This indifference to death and basic sanitation did not stop at pack animals, but also applied to humans. Wounded men were being abandoned in dug-outs or left in no man's land, and the hospitals remained overcrowded and messy after the casualties suffered from 9 to 11 May. Even the *chasseurs* fell into a relative

---

[34] Ibid.

[35] SHD, 24N1991; «Le Général Pétain, Commandant le 33ᵉ *C.A. à le Général Commandant l'Artillerie du C.A. (pour exécution); le Général Commandant le 18ᵉ, 70ᵉ, 77ᵉ Division pour avis*» 17 *mai 1915*.

[36] Ibid.

[37] Ibid. «Note de Service» 33ᵉ *Corps d'Armée, 20 mai 1915*.

[38] Ibid.

[39] Ibid. «Le Colonel *Picat commandant l'Artillerie de la 77 Division à Monsieur le Général commandant la 77ᵉ Division» 22 mai 1915*.

[40] Ibid. «Note de Service 77th Division à A.D. 77» 19 *mai 1915*.

torpor, and were reprimanded for their unwillingness to do their full duty to care for fallen soldiers.[41]

As important as it was to get the artillery into a position from which it could most effectively engage the enemy, the reconstruction of a solid defensive network and adjustment to the changing character of the division (the temporary addition of 67[e] Bde and the presence of a new divisional commander) were much more urgent. The primary tasks for 77e DI from 18 to 25 May would be to continue improvement of the forward trenches, connecting these trenches to the rear with new *boyaux*, and the digging of new deep dug-outs in the front lines (a task started on 13 May).[42] By 19 May Pétain was already anxious that the work at the front was not progressing as quickly as it might. He sent a note to the entire corps, complaining about the lack of progress and stating that reinforcements needed to take part in the reconstruction of XXXIII CA's positions as well.[43] The divisional commanders were all berated for not ensuring that the work assigned was being done satisfactorily, and they were pushed to achieve results. Pétain suggested that they use every means at their disposal to get front-line troops to complete the tasks assigned to them, even to the point of refusing to relieve units until they had completed their assigned work.[44] He demanded daily reports on the work done and exhorted divisional commanders to make daily inspections to ensure that things were progressing as they should.

Pétain's displeasure met with immediate and appropriate response. Pillot ordered that the territorial troops assigned to assist the front-line units with their work would be under the direct command of the battalion they were assisting in order to minimize any friction and discord which the lack of clear command authority might have allowed.[45] Pillot channelled some of Pétain's ire, and condemned the work the territorials had done up to that point as having been 'utterly insufficient'. He demanded that daily reports be delivered to the engineers in charge of overseeing the digging. The engineers would then collate the various reports coming in, and pass them on to divisional HQ. Pillot also had to sort out bureaucratic turf wars between the infantry and the engineers which were being exacerbated by a shortage of materiel. Frequently, when infantry units in the 77e required specific tools or resupply to replace tools lost or broken, they would ask the engineers, who would roundly refuse any request made by the infantry – a problem touched upon in Pétain's 19 May note.[46] Inevitably the request would

41 SHD, 24N1991; «77e *Division à Groupe Chasseurs [et] Médecin Divisionnaire*» 18 *mai 1915*.

42 Ibid. «*Travaux à Exécuter dans le Secteur de la Division*» *17 mai 1915*.

43 Ibid. «*77e Division à 67e et 88e Brigades, 159e et 83e R.I. (directement), Gr. Chasseurs, – Génie, 25e T – 6e Btn du 231 R.I.*» *19 mai 1915*.

44 Ibid.

45 SHD, 24N1991. «*77e Division à 231e R.I., 25e T., Génie Divisionnaire*» 19 *mai 1915*.

46 Ibid. «*77e Division à 67e et 88e Brigades, 159e et 83e R.I. (directement), Gr. Chasseurs, Génie, 25e T – 6e Btn du 231 R.I.*» *19 mai 1915*.

land on Pillot's desk, and he would be required to find the tools necessary for the infantry to carry on with their assigned work. This not only greatly expanded the bureaucratic workload of Pillot's fresh command (thus complicating an already difficult transfer of power), but also slowed down productivity in the division, and increased suspicion and antagonism between the various arms – hardly a beneficial development. On 20 May Pillot addressed the issue by ordering that all requests for tools by the infantry were to be directed to the engineers, who would fulfil them without question, and that the divisional commander should only be informed if the engineers were unable to do so.[47] Thus the engineers were kept in check, and the division could more smoothly continue its arduous excavation before Souchez and Vimy Ridge.

On 25 May, after a fortnight of work, rest and reorganization, elements of Tenth Army (IX, XXI and XXXIII CAs) were ready to make an attempt to capture Souchez and other problematic enemy strong-points in preparation for a later drive on Vimy Ridge. The situation on 77ᵉ DI's front was still not ideal. There were not as many jumping-off trenches as there had been for 9 May, and the entire network, especially the *boyaux*, was insufficiently developed. This slowed down the circulation of troops and supplies, and invited more casualties as men lacked decent protection from enemy fire. Incessant German artillery bombardments greatly slowed the French consolidation of their captured German trenches, and forced most of the work to be done during the short spring nights.[48] Attempts were made to encourage the men to continue to work during the day, but even Pillot had to admit that doing so only resulted in more casualties. He ultimately gave the somewhat contradictory order to continue work during the day, but to do so with only a minimum of personnel to reduce the number of casualties the division might suffer.[49]

The objectives for 77ᵉ DI on 25 May were the Château Carleul, the *ilôt de Souchez* (a small, detached set of buildings) and Souchez cemetery (Figure 5.1).[50] A complex artillery bombardment with repeated starts and stops designed to confuse the Germans was orchestrated by Pillot; Delmotte (67ᵉ Bde) was to be in command of the attack. 37mm cannons were positioned to inflict enfilade fire on the Neuville–Souchez road; the 58mm cannons were concentrated on the Souchez cemetery and the first few houses north of it.[51] A two-hour long artillery preparation from 10.40 to 12.40 would precede the attack. At 12.40 the artillery firing on the first German line would roll forward to the second line at 25m bounds.[52] Between 30

---

47 Ibid. «77ᵉ *Division à 67ᵉ et 88ᵉ Brigades, Gr. Chasseurs, Génie Divis., Cdt 60ᵉ Btn chase.» 20 mai 1915.*

48 SHD, 24N1991; «Le Commandant *du 1ᵉ Bataillon du 25ᵉ Régt. Terr. à 77ᵉ Division» 24 mai 1915.*

49 Ibid. «Le Général Pillot Commandant la 77ᵉ *Division à Monsieur le Commandant du Génie Divisionnaire»* 24 *mai 1915.*

50 Ibid. «*Ordre d'Attaque pour la Journée du 25 Mai» 25 mai 1915, 02.00.*

51 Ibid.

52 Ibid.

Figure 5.1 Souchez, with Château Carleul (upper left) and Souchez cemetery (lower middle)

Source: SHD, 24N1991; «Le Commandant *du 1e Bataillon du 25e Régt. Terr. à 77e Division» 24 mai 1915*. Unnamed map.

and 50 minutes after this initial barrage a second rolling barrage would begin with different batteries firing individual barrages which would roll towards a variety of targets (the barrages were staggered within this 20-minute time frame based on the target that a given barrage was meant to reach). The hope was to ensure not only surprise, but also to provide individualized support to the different assault units as they approached their tertiary objectives, and to allow for flexibility if the infantry were held up or if the situation changed in any way. It was a logical response to the fact that 77e DI would not be advancing in a simple straight line, but would be required to fan out towards different objectives.

When the infantry went over the top at 12.40 they met with mixed results. One battalion of 97e RI (the other two were in reserve) captured its primary objective – the trenches between the Neuville–Souchez road and the *route de Bethune* – with little trouble, and began the process of *nettoyage* and preparing for a second effort on its ulterior objectives.[53] To its left, 14e RI (67e Bde) found the artillery preparation insufficient upon leaving its trenches, and was met with intense German fire. While elements of 14e RI's left were able to advance and capture the first line of German trenches, the rest of the regiment was compelled to quickly return to the safety of its *parallèles de départ*.[54] With its left exposed, 97e RI came under increasing enfilade fire, and was forced to fall back and dig in in no man's land, roughly 50m from the German trenches.[55] A second attack was organized at 18.00 in the hope of getting 14e RI to advance. The artillery preparation was similarly complex, although it rose to a crescendo just before the attack, which may have betrayed any attempt at surprise; again the rolling barrage was to advance at 25m intervals. Despite these further attempts, 14e RI was unable to make any headway against the German defences. As a result, those forward troops which had made progress that day were required to pull back, which they did with heavy losses: the two assault companies of the 97e suffered over 50 per cent casualties that day.[56]

On 26 May another attempt was made, but met with only very limited gains despite some fairly creative infantry tactics (grenadiers were sent on the wings ahead of the main attack to enfilade the German trenches with grenades while the rest of the infantry advanced across no man's land).[57] Undoubtedly frustrated, Pétain wrote to d'Urbal the next day to describe the situation at the front. He stated that while before 9 May there was very little German activity and the French were able to achieve surprise, the situation had radically changed. The Germans were inflicting accurate and intense daily bombardments on the French, which made any offensive action all but impossible.[58] Measures to help the infantry advance – such as varying the duration, intensity and pattern of artillery bombardments – had failed to make any appreciable difference as the enemy artillery could still be called into action at a moment's notice.

The real problem, then, was how to engage the enemy artillery in effective counter-battery fire. Most of the German guns were pre-registering on key areas through which the attacking French infantry would have to pass. They could therefore remain silent until the moment of attack, at which point they would unleash a violent counter-barrage, which more often than not sent the attacking

---

53 SHD, 26N407/2; 77e DI JMO.

54 Ibid.

55 Ibid.

56 Ibid.

57 SHD, 24N1991; «*Ordre d'Attaque du 14e R.I. pour la journée 26 Mai*».

58 SHD, 24N1991; «Le Général Pétain, Commandant le 33e *C.A. à Monsieur le Général Commandant la 10e Armée*» *27 mai 1915*.

poilus scrambling back into their *parallèles de départ*.[59] As a result, counter-battery fire largely consisted of waiting for the enemy guns to open up, and then lobbing shells in the general direction of the enemy batteries and hoping for the best. The results of this desperate attempt to suppress the enemy artillery were frequently negligible.[60] The answer, Pétain felt, was better aerial reconnaissance. Aircraft were incapable of correcting artillery fire during a battle owing to the general confusion and the limited range of their wireless transmitters.[61] Even if they could, the process of registering a battery on a new target was sufficiently time-consuming that the enemy battery in question could easily fall quiet or relocate, thus thwarting the French effort.

Pétain believed that the only answer would be to make detailed maps of the enemy rear areas, and systematically fire on their artillery emplacements over a long period of time instead of hoping to stifle the German guns all at once on the day of the attack.[62] The squadron commander for XXXIII CA had informed Pétain that in an eight-hour period his squadron could assist 15–20 batteries (thus, roughly 60–80 guns) via wireless in their registration against enemy targets.[63] This limitation further supported Pétain's push for increased artillery pre-planning to match the German methods, which had proved so effective since 11 May. Pétain envisioned extensive maps being drawn up and distributed among various batteries, with zones for counter-battery fire being assigned to either 75mm guns or the heavy artillery based on the distance to the target, and felt that the Tenth Army artillery commander should take charge in organizing the effort.[64]

Unfortunately, Pétain's recommendations were not taken up, and Tenth Army would have a difficult time responding to the rapidly evolving German defensive tactics. On 27 May, the same day that Pétain submitted his ideas to d'Urbal, the Moroccan Division re-entered the line for the first time since 11 May.[65] It took over part of 77e DI's sector, relieving both 97e and 159e RIs (88e Bde) and temporarily taking command of 67e Bde. 67e Bde would remain in the line until replaced by the returning 77^e^ DI, soon after which the 67^e^ Bde would finally be returned to its 34^e^ DI. From 11 to 27 May 77^e^ DI had not accomplished much. It had successfully connected the trenches it had captured on 9 May to its old front line, but had not done so in totality before coming out of the line. Some small–medium-scale attacks had been launched, but none of them bore fruit. It was hardly the result one would have expected from *La Division Barbot*.

---

59 Ibid.

60 Ibid.

61 Ibid.

62 Ibid.

63 Ibid.

64 SHD, 24N1991; «Le Général Pétain, Commandant le 33e *C.A. à Monsieur le Général Commandant la 10e Armée*» *27 mai 1915*.

65 SHD, 26N407/2; 77e DI JMO.

Back to Basics: 34e DI

The situation of 34e DI after 11 May was fundamentally different from that of 77e DI. Nevertheless, this did not preclude the division from having to deal with many of the same issues which dogged units throughout Tenth Army in the days and weeks following 9 May. 34e DI had made only the most nominal advance during the opening phase of Second Artois, which made the process of linking up its new front line with the rear significantly easier. The short advance also meant that the divisional artillery (*artillerie de division*, AD/34) would not have to relocate closer to the front, and thus could continue to fire on German positions which its guns had already registered upon in the preparation for 9 May. Still, the intense bombardment suffered by 34e DI did necessitate a fair amount of work repairing battered trenches, and the division would have to recover from the heavy casualties suffered on 9 May, both physically and mentally; morale would dip frighteningly low in the weeks after 11 May. How the division weathered the storm, learned from its mistakes and eventually reached a level of competence is the story of how much good leadership and a conscious effort to improve can affect a formation. It also reflects the fact that even the most battered fighting force can spring back if given the right circumstances, an interesting metaphor for the French army as a whole as it developed from the army that performed so poorly in 1914 to the army that would dominate operations on the Western Front from mid-1916 onwards.

As with 77e DI, 34e DI was also required to make partial offensives in the days after 12 May which, in theory, would secure the flanks of XX and XXXIII CAs. Urged on by over-optimistic messages from Tenth Army, which proclaimed that the Germans had been morally weakened, asked to carry out difficult supportive attacks of dubious value, and deep in a morale crisis, the quality of the attacks launched by 34e DI plummeted.[66] Already by 12 May the division had fallen into its old bad habits from First Champagne, with the order to attack that day simply being, 'same objectives, and the same disposition as the attacks of the last few days'.[67] Artillery fire was to continue over the course of the night of 12–13 May, with the attack going in at first light (roughly 03.00). A fixed barrage would be fired on the second German position after the first had been captured, to give the men time to consolidate their gains.[68] The attack was launched according to plan; it was not successful.

In the hope of continuing to support the main axis of attack a combined offensive between 33e and 34e DIs was planned for 15 May. A note from Joffre stating his confidence that XVII CA would capture Thélus on 15 May was circulated in the

[66] SHD, 24N741; «*Ordre Général d'Opérations No. 303 pour la journée du 12 Mai» 11 mai 1915*.

[67] Ibid. «*Ordre Général d'Operations*» 12 *mai 1915*: «Elle aura les *mêmes objectifs, et le même dispositif que les attaques des jours précédents*.»

[68] Ibid.

hopes of raising morale, although it is doubtful that it had any positive effect.[69] A second note, sent from XVII CA headquarters, detailed the lessons learned by XVII CA since Champagne. It cast meticulous and well-controlled artillery fire, along with close liaison between the infantry and artillery, as the primary components of success.[70] The preparatory bombardment, organized by Dumas, was to be a mere six minutes long, but for the first positively documented time in Second Artois 34ᵉ DI would use a rolling barrage.[71] When the attack went in the advancing *poilus* were met with a violent German counter-barrage which could not be crossed, thus forcing XVII CA's infantry back into their trenches. Despite a fair amount of counter-battery fire the French guns seemed to have had no effect on their German counterparts, whose deadly fire made any advance impossible.

34ᵉ DI's repeated failures from 9 to 15 May did not facilitate any positive feelings between de Lobit and his superiors. When questioned as to why the attack on 15 May had failed, de Lobit responded emphatically that the infantry and artillery had acted according to the prescriptions laid out by XVII CA and that 'no one' was to blame for their failure.[72] De Lobit insisted that not only was the artillery preparation ineffective, but the plan itself was also fundamentally flawed, and the Germans knew exactly when the attack was coming. Amid complaints from Tenth Army about the pitiful results XVII CA had achieved, Dumas chose to take 34ᵉ DI close in hand; over the weeks that followed he would make extensive efforts to reform 34ᵉ DI and get it to a point of basic competence.

He started on 16 May by sending two notes to 34ᵉ DI boiling over with his frustration and annoyance. Dumas wrote to complain bitterly about the excessive filth of the 34ᵉ DI front, especially the lack of disinfected latrines.[73] He stated that officers needed to make regular rounds and demand cleanliness on threat of punishment. This was especially important as warm weather was approaching, which would raise the risk of an infectious outbreak. In his second letter, Dumas wrote to condemn the actions of certain lightly injured soldiers who had taken to sitting atop the hill at St Catherine to watch the artillery bombardments and cheer on French efforts, thus revealing the presence of XVII CA's field hospital in defilade behind St Catherine.[74] This activity inevitably caught the attention of

---

69 Ibid. *«Le Général J.B. Dumas Comm'at le 17ᵉ Corps d'Armée aux Généraux Commt. les 33ᵉ et 34ᵉ divisions» 14 mai 1915.*

70 SHD, 24N741; «Instruction de Rappel *sur la Méthode de Travail Adoptée par Expérience au 17ᵉ Corps Depuis la Début des Operations en Champagne» 15 mai 1915.*

71 Ibid. «Le Général Commandant le 17ᵉ *Corps aux Généraux commandant les 33 et 34 Divisions» 15 mai 1915.*

72 Ibid. «Le Général de Lobit Commandant la 34ᵉ *Division d'Infanterie à M. le Général Commandant le 17ᵉ Corps d'Armée»* 16 *mai 1915.*

73 Ibid. «Le Général J.B. Dumas, Commandant le 17ᵉ *Corps d'Armée, au Général Commandant le 34ᵉ Division» 16 mai 1915.*

74 SHD, 24N741; «Note de Service» 16 *mai 1915.*

German gunners, who would bombard the area – which made the running of a hospital considerably more difficult.

On 20 May XVII CA would be relieved by X CA, and two days later be relocated to a position on the far right of Tenth Army, which included the northern reaches of Arras.[75] It was here, in the quietest sector of Second Artois, that Dumas would seriously begin to reform 34e DI. During the transfer Dumas set the tone by ordering a much more thorough documentation of maps, panoramic sketches, and reports, which would be circulated among the infantry, artillery and engineers.[76] Daily reports of action, and any observations, were to be written by the various branches and units within XVII CA, and circulated throughout. Likewise, aerial reconnaissance reports from the corps' air squadron would be circulated to lower-level units to increase the flow of information.

Once in position, XVII CA began the process of overhauling the dilapidated trench network it had inherited. *Boyaux* had to be enlarged to protect troops during a heavy bombardment; dug-outs and latrines needed to be built; trench guns needed to be quickly installed and begin acquainting themselves with their new targets. Interlocking fields of fire needed to be designed for the corps' machine guns, and the fresh spring grass, a serious threat to observation from the trenches, needed regular trimming.[77] On the corps' second day in their new sector Dumas took a tour of his new position; the situation was far from ideal. The front trenches were situated in such a way that they were not in line of sight with the rear, making visual communication impossible.[78] Dumas ordered that the situation be rectified by re-siting the forward trenches, and assigned men the job of maintaining a constant liaison between the front and the rear. The listening posts were also poorly sited in positions beneath the German trenches opposite them. They offered no observation of the enemy trenches safely positioned atop a small rise.[79] To make matters worse, the trenches were full of faecal matter and other human debris which needed to be cleared up before they could spread disease.

While inspecting the trenches, Dumas came across men assigned to guard duty sleeping in broad daylight, free from any scrutiny from their superiors (who were mostly to be found resting in their command posts).[80] Disgusted, Dumas ordered a much tighter control of guard duty, plus the commission of snipers

---

75 SHD, 26N163/3; XVII CA JMO.

76 SHD, 24N741; «Le Général J.B. Dumas, Commandant le 17e *Corps d'Armée au Général Commandant la 33e D.I., au Général Commandant la 34e D.I., au Général Commandant la 88e D.T., au Colonel Commandant l'Artillerie du 17e C.A., au Colonel Commandant le Génie du 17e C.A.» 20 mai 1915.*

77 Ibid. «Le Général J.B. Dumas, Commandant le 17e *Corps d'Armée au Général Commandant la 34e Division d'Infanterie*» 23 *mai 1915.*

78 Ibid. «Observations du Général Commandant le 17e *Corps d'Armée au Cours de sa Visite des Tranchées le 24 Mai 1915».*

79 Ibid.

80 Ibid.

and the launching of regular patrols into no man's land. Reports were to be sent daily by divisional commanders to Dumas to ensure that his measures were being followed. Three days later, on 26 May, Dumas took another tour of the front lines – and found the situation little improved. The trenches were still filthy; latrines were dilapidated and had not been disinfected.[81] The risk of infection was increased by the excessive concentration of men in the front trenches. This not only made sanitation difficult, but also increased the division's exposure to casualties in the face of any aggressive enemy action. Most objectionable to Dumas, men he interrogated at random did not know who their corps commander was – and some were ignorant of who their regiment or even battalion commanders were.[82] Dumas berated his subordinates for allowing the martial spirit of the men to fall so low, and thus took on one more facet of his corps to train and rebuild.

In addition to the practical and cultural reorganization that was needed in 34e DI, Dumas would also take pains to prepare his subordinates intellectually. On 25 May, the same day 77e DI made its failed attack on Souchez and the Château Carleul, Dumas requested that de Lobit plan 34e DI's attack on Chantecler. Successive drafts of the attack would be sent back to Dumas for editing, thus training de Lobit to lead his division better. Dumas listed a long series of variables for de Lobit to specify, including the enemy trenches which needed neutralizing (versus those that needed outright destruction by heavy artillery), and the artillery required to meet these goals. De Lobit was also ordered to specify the trench work needed before the attack; the formation of the assault and support waves; the specific objectives of each unit; and the support required from neighbouring units to ensure a successful attack.[83]

Two days later de Lobit submitted his initial plan to Dumas. In his plan he recognized the huge importance of a visibly strong and well-organized preparation, not only for the necessary softening of the enemy defences but also for the effect it would have on the morale of his troops.[84] To help ensure a strong preparation de Lobit asked for a serious augmentation of his artillery capacity, in the form of more and better shells (he had largely shrapnel shells) and more de Bange 155Cs (although modern 155CTRs would be preferred if they were available), or at the very least some additional 220mm mortars to help make up for his deficiency in heavy artillery. De Lobit addressed all of the particulars Dumas had raised, including how many shells he thought would be needed (roughly 25,000 75mm shells and 4,500 heavy shells, with stated quantities spent on specific targets along

---

[81] Ibid. «Observations du Général J.B. Dumas à la suite de *sa visite du 26 Mai 1915 dans les tranchées*» 26 *mai 1915*.

[82] Ibid.

[83] SHD, 24N741; «Le Général J.B. Dumas, *Commandant le 17e Corps d'Armée au Général Commandant la 34e Division» 25 mai 1915*.

[84] Ibid. «Le *Général de Lobit Cdt la 34 D.I. à Monsieur le Général Cdt le 17e C.A.» 27 mai 1915*.

the front), in addition to detailing the infantry formation and organization.[85] On 28 May a more prosaic set of instructions was written out, which placed the attack in its wider context and assigned detailed roles to the engineers, including the handling of explosives and repairing any captured trenches.

The next day Dumas sent his reply to de Lobit. While the plan seemed all right it was not perfect, and a few key steps had been skipped. De Lobit had taken the time to carefully draw up the formation of the attacking infantry, Dumas explained, but he had failed to give each company a specific objective, the lack of which would undoubtedly cause confusion during the attack.[86] Dumas also felt that de Lobit had spread his AL far too thinly, which would result in an incomplete preparation. He suggested that what munitions were available be focused not on 'areas', but on singular, discrete targets such as machine guns and certain flanking positions. The field artillery would be left to make blanket bombardments of different areas to impede any German repair work.[87] This would not only ensure that the AL fired on important pieces of the German defensive system instead of wasting munitions on general-purpose trench-works, but would also ensure that a greater quantity of munitions would be spent on each individual target.

This assessment was supported by the commander of the squadron attached to XVII CA's 3^e AL battery. He reported that while the artillery bombardment had been well registered, the 160 shells assigned to 3^e battery that day would be wholly insufficient for the task before them.[88] In response, de Lobit reorganized the artillery bombardment to focus more on the most crucial segments of the forward defences opposite him, and sent an appeal for four extra heavy guns, plus an extra 100 rounds for each gun, for use the next day, 2 June.[89] As per Dumas' recommendation the field artillery was told to concentrate on supporting the efforts of the heavy artillery first and foremost, while the 220mm mortars were focused increasingly on the salient protecting Chantecler along the main axis of the upcoming attack.[90] At the same time de Lobit was trying to convince the mortar teams of neighbouring units to also fire on Chantecler in support of 34^e DI's efforts.

Despite serious efforts by 34^e DI to improve its situation it still occupied trenches below their German counterparts, and tall grass continued to obstruct terrestrial observation. As a result, the only available source of information was aerial reconnaissance, which was not as accurate as one might have hoped. Early aerial sorties could make out the beginnings of a few key dents in the enemy wire,

---

85 Ibid.

86 Ibid. «Le Général J.B. Dumas, *Commandant le 17e Corps d'Armée au Général Commandant la 34e Division» 29 mai 1915.*

87 Ibid.

88 SHD, 24N741; «Au *sujet des tirs du 3e Groupement d'A.L. (Chef d'escadron Delxxx)» 1er juin 1915.*

89 Ibid. «Au *sujet tirs d'Artillerie lourde»* 1er *juin 1915.*

90 Ibid. «Au *sujet des tirs du 3e Groupement d'A.L. (Chef d'escadron Delxxx)» 1er juin 1915.*

but could not confirm whether they constituted open breaches. Still, early reports were confident that the bombardment was making progress.[91] This was sharply countered on 2 June when it was reported by XVII CA's air squadron that the bombardment had made no real impact on the German defences at all. Attempts were made by de Lobit to enhance aircraft–infantry–artillery liaison, but it did little to affect what was a very difficult position.

Despite the many challenges 34ᵉ DI faced while preparing for its attack on Chantecler there was at least some reassuring news: 67ᵉ Bde would return to its native 34ᵉ DI in the first days of June, thus bringing the division back up to full strength (technically). Aerial reconnaissance had reported that as the German trenches defending Chantecler were sited on top of a rise, as opposed to on the forward slope, they would not have very good observation of the field over which 34ᵉ DI would have to advance.[92] This observation would be further reduced by the heavy smoke that the final bombardment would produce, which would linger over the German trenches if the wind stayed as calm as it had been in previous days. It was estimated that the French would be able to reach the German wire before being noticed.[93] Furthermore, the German trench network was poorly organized in front of 34ᵉ DI, with many areas seemingly not in visual contact with each other, which would make it excessively difficult for the Germans to respond quickly to a French attack.

With hard work and sound leadership the condition of 34ᵉ DI's trenches had slowly begun to improve. An inspection on 3 June found that work had been progressing nicely, a verdict repeated in a note by Dumas issued a few days later that, while still providing numerous recommendations for improvement, exhibited a dramatic change in his tone from one of wrathful disgust to one of calm and relative confidence.[94] His tutelage had brought 34ᵉ DI a long way in a remarkably short period of time and de Lobit seemed to be handling his very difficult preparation with an ever-increasing professionalism.

The skills de Lobit had learned and honed since mid-May would be pushed to their limit in the days leading up to the attack on Chantecler. The German position above the French trenches not only made observation a problem, but even simple bombardments were proving difficult. 34ᵉ DI did not enjoy access to many high-arcing guns, and what models it did have were more often than not 30-year-old de Bange 155Cs. As a result, the AD/34 (*artillerie de division*) – already finding it difficult to make a serious impact on enemy defences owing to the relative weakness of 75mm shells – had to rely on dubious methodology,

---

91 Ibid. «*Compte-rendu, observateur aux tranchées = Dubois (3ᵉ Groupe, 23ᵉ artillerie)» 1ᵉʳ juin 1915.*

92 SHD, 24N741; «Rapport du lieutenant *observateur Dubois, 3ᵉ Groupe, 23ᵉ artillerie*» 3 *juin 1915.*

93 Ibid.

94 Ibid. «Rapport du Lieutenant Vidal du *détachement Télégraphique du 17ᵉᵐᵉ Corps d'Armée, sur les lignes télégraphiques d'Infanterie de la 34ᵉ Division*».

including the *Plaquette Malandrin*, to bombard the German trenches.[95] From early on a heavy emphasis was laid on the infantry and aircraft doing whatever they could to help locate enemy batteries owing to 34e DI's poor terrestrial observation capabilities in its new sector, but the results were not particularly impressive; and a shortage of trained signallers did little to help the situation.[96] The bombardment by mortars and de Bange guns was made even less effective by the German ability to do significant repair work at night as the division's shell shortage made their disruptive fire on the German trenches exceedingly weak. Field artillery were ordered to watch their fire, shell by shell, to ensure that no munitions were wasted, and heavy batteries were often restricted to only 100 shells per gun per day (1,200 total heavy shells daily for the division).[97]

The problem posed by 34e DI's limited number of shells was exacerbated by their being the wrong type of shell. Around 6 or 7 June the 3e AC group wrote to de Lobit to explain their situation. The group had already lost enough 75s to reduce the number of guns per battery from four to three apiece.[98] To make matters worse, they lacked percussion shells, and instead had to rely on timed fuses which, owing to their variability, represented a significantly less efficient shell. As the precise time of explosion could not be guaranteed, shells would very often explode at less than ideal times, which frequently left bombardments incomplete despite having fired a considerable number of shells. As the more efficient percussion shells, which would explode on impact, were few in number, larger quantities of timed-fuse shells were having to be committed to the preparatory bombardment, which denuded the artillery of shells for interdiction fire overnight, allowing the Germans to repair whatever damage had been done during the day.[99] The result of this situation is obvious. As a stop-gap measure, the artillery suggested that the infantry do more to impede German repair efforts by rifle and machine-gun fire (perhaps even raids). The document states that this should be considered one of the primary responsibilities of the infantry in any case, since they could disrupt German efforts to repair their wire emplacements much more cheaply and effectively than artillery could.[100] This, unfortunately, was not only ineffective, but would also result in increased infantry casualties.

Problems of increasing complexity, such as those posed by the division's material paucity, continued to arise and were dealt with by de Lobit as well as

---

95 SHD, 24N741; «*Compte-rendu au sujet de la mission de l'A.D. 34» 24 mai 1915.*

96 Ibid. «Le Général J.B. Dumas, Commandant le 17e *Corps d'Armée: au Général Commandant la 33e D.I., au Général Commandant la 34e D.I., au Général Commandant la 88e D.T., au Colonel Commandant l'Artillerie du 17e C.A., au Capitaine Commandant l'Escadrille C-56» 23 mai 1915.*

97 Ibid. «Programme *Général des tirs d'artillerie pour la journée du 4 juin*» 3 *juin 1915.*

98 Ibid. «Rapport du *Tir*» 23e *Artillerie, 3e Groupe.*

99 Ibid.

100 Ibid.

possible. On 8 June 34ᵉ DI lost four 75s: three suffered fatal barrel-wear (swelling) due to over-use and the fourth exploded.[101] This brought down the total number of guns in the division to 29 and inflamed the consternation of Colonel Malet, commanding the AD/34. He wrote to de Lobit on 8 June and informed him that the guns of 34ᵉ DI were firing more than 200 shells each day, thus putting dangerous levels of stress on already worn guns. Malet warned that if the situation continued as it was they might face stark losses before the attack was ever launched, and at the very least the tired guns would have difficulty keeping up with even the most modest firing plan on the day of the attack.[102] De Lobit responded the next day, immediately after learning that yet another gun had exploded, and reduced the per-gun quota to only 50 shells a day. He would furthermore ask Dumas if he could find replacements for his lost guns, but none were forthcoming.

The situation would continue to deteriorate as 34ᵉ DI began in earnest to prepare for the attack on Chantecler. By 12 June the division had lost a further two guns, while six were operating at a significantly reduced capacity (four could fire only shrapnel shells without undue damage to the guns and two fired very irregularly).[103] This, then, reduced the number of fully operable field guns in 34ᵉ DI to just 20: less than two-thirds of their full-strength contingent of 36 field guns, and fewer than the number of planned breaches to be made in the enemy wire. The most intense period of artillery preparation would begin the next day, on 13 June, and threatened to leave the division without any serviceable artillery come the 16th. De Lobit's regular and increasingly frustrated demands for new guns went unfulfilled: the only communication from Tenth Army was a noncommittal statement that they would be supplied with all the munitions they would need.

The infantry of 34ᵉ DI also presented its share of difficulties. Despite repeated efforts at achieving even a basic level of sanitation, the men of 34ᵉ DI still used *boyaux* and saps as latrines, and refused to disinfect their waste.[104] Water was not being distributed carefully enough (being supplied in open containers instead of closed casks with taps), exposing the men to an increased risk of infection; and telephone lines were frequently neglected and fell into disrepair. The sector, despite three weeks of work, was still far from ready. Company command posts were located too far forward, which made lateral liaison difficult and overly exposed company commanders to German artillery bombardments; and the trenches were still in need of much improvement.[105] The fact that the only regularly positive

---

[101] SHD, 24N741; «Le Colonel *Malet, Cdt l'AD.34 à M. le Général Cdt la 34ᵉ Division» 9 juin 1915.*

[102] Ibid.

[103] Ibid. «Le *Général de Lobit Comt la 34 DI à M. le Général Cdt la 17ᵉ CA» 12 juin 1915.*

[104] SHD, 24N741; «Le Général J.B. Dumas, Commandant le 17ᵉ *Corps d'Armée au Général Commandant la 34ᵉ Division d'Infanterie»* 11 *juin 1915.*

[105] Ibid. «Observations *faites par le général Ct la 34ᵉ Division au cours de son inspection du secteur dans la matinée du 12 Juin»* 13 *juin 1915.*

comments made about the division concerned its strong and well-sited wire emplacements should provide some insight into the mindset of the infantry.[106]

The men of 34e DI, who did not know who their corps commander was even when he stood before them, had not gained any real martial spirit in the three weeks they spent digging trenches on the far-off, forgotten right wing of Tenth Army. De Lobit found men often leaving their weapons in their dug-outs instead of carrying them. Listening post duty was still not taken seriously, with men often falling asleep or failing to be present at all. Furthermore, divisional mortars were neglected, with only two out of 16 seeing regular use despite their being the perfect weapon for the very difficult preparatory bombardment that 34e DI was attempting to orchestrate.[107] Symptomatic of the malaise that had shrouded the division since 9 May, 59e RI (68e Bde) even failed to organize all its men when they were to be relieved. As a result, many men who had been asleep at the time awoke to find that the regiment had marched back to the rear without them, while other individual soldiers had apparently made their way to the rear in their own time (and well before they were supposed to have done so).[108]

Despite his many shortcomings, there is no doubt that the leadership skills of de Lobit had grown rapidly in the weeks since the debacle of 9–11 May. His plan of attack against Chantecler was every bit as detailed and thoughtful as one could expect. Under the close tutelage of Dumas, de Lobit began, more and more, to lead his division properly, and took to demanding ever tighter liaison with his subordinates and making first-hand inspections of his front. While his situation was far from ideal, and the Germans maintained a serious advantage owing to their position on high ground above the French trenches, he did make 34e DI significantly more efficient than it had been at the start of Second Artois. Whether or not his work would be sufficient to overcome a remarkably lax ideology among the men would be proven when Tenth Army launched its second major bid at Vimy Ridge on 16 June.

---

106 Ibid. «*Compte rendu*» 67*e Bde, 14 juin 1915.*

107 SHD, 24N741; «Observations *faites par le général Ct la 34e Division au cours de son inspection du secteur dans la matinée du 12 Juin*» 13 *juin 1915.*

108 Ibid.

Chapter 6
The Second General Offensive (13–18 June)

After the opening of Second Artois on 9 May 1915 it would take Tenth Army five weeks before it was ready to make a second bid for Vimy Ridge. In that five weeks reinforcements were brought up; attacks continued to inch Tenth Army nearer its objectives; new trenches were dug and old ones improved. On paper Tenth Army seems to have had every reason to expect success. Unfortunately, Tenth Army was not progressing in a vacuum, but was competing against an opponent who had received a bloody nose on 9 May, and would work very hard to ensure that it did not happen again. While *Grand Quartier Général* (GQG) re-evaluated the doctrine laid out in Note 5779 based on the experiences of Tenth Army in early May, the new lessons did not have time to make it to the front, and troops there were too busy with day-to-day operations to perform much training. The Germans, on the other hand, could make their position impregnable without much work or too drastic a doctrinal overhaul; this was especially true in its vastly increased use of pre-sited artillery barrages to counter French infantry assaults. As a result, Tenth Army would find itself somewhat stronger and in a slightly better position than it had been on 9 May, but faced with a much-improved German defensive system with predictable results.

The Build-up to 16 June

On 16 June 1915 the French Tenth Army launched the largest single trench offensive on the Western Front to date: 799 field guns and 355 heavies had been amassed for what was to be the final victorious push on Vimy Ridge. Tenth Army had received substantial reinforcements, yet the end result did not amount to many more guns than had been present on 9 May. Likewise, infantry reinforcements had poured in, but only managed to get infantry levels up to slightly above where they had been five weeks earlier. Extra machine guns and 58mm cannons had been begged or borrowed from neighbouring armies, often leaving their defences perilously thin, in order to give Tenth Army's attack as much weight as possible.[1]

Where Tenth Army could boast a significant increase over its 9 May strength was in munitions. According to the French Official History, Tenth Army would fire 497,122 shells of all calibres (417,780 field-gun shells and 79,342 heavy shells)

[1] *AFGG, Tome III*, p. 79.

from 10 to 16 June plus an additional 221,429 (192,905 field, 28,524 heavy) from 17 to 18 June; between 3 May and 9 May Tenth Army fired only 265,430 (222,485 field, 42,945 heavy).[2] Another important area of improvement would be the overall scale of operations. On 9 May the French and British attacked along a continuous front of nearly 25km. In mid-June the French Army would be launching three major supporting attacks (one by Castelnau's Second Army from 7 to 13 June, another by Dubois' Sixth Army from 6 to 16 June and a third by Maud'huy's Seventh Army from 14 to 22 June) in addition to the main push in Artois, which would again be supported by a British attack, this time focused on Zillebeke (thus not attacking on a contiguous front with the French as they had done on 9 May).[3] Artillery preparation was to begin on 13 June, with most corps attacking in the same general direction they had on 9 May: XXI CA was to advance beyond Notre Dame de Lorette (NDL) into the *Bois de Givenchy*; XX CA was to finish taking Neuville and the *Labyrinthe*. A major change, however, occurred along the main axis of the attack. XXXIII CA was shifted slightly to the left, and ordered to throw its weight against Souchez (including the Château Carleul), *cote* 119 and, ultimately, Givenchy, while IX CA was to have the task of storming Vimy Ridge.

IX CA, commanded by General Curé, had had a largely auxiliary role in the 9 May attack, merely protecting the flank of XXI CA on their right and maintaining liaison with the British on their left. On the night of 29–30 May the relatively fresh corps began the final phases of relocating from the left wing of Tenth Army into the heart of the attack.[4] The corps did receive some training during the transition, especially in the use of grenades, but only on a limited scale as the need to press on digging trenches and preparing for the offensive took priority. Before 16 June IX CA participated in a few small- to medium-scale offensives against Neuville in conjunction with XX CA, which would act as practice runs before the general offensive. IX CA did not perform well. One such attack on June 10 was quickly aborted after the lead units went to ground under enemy fire and refused to press on; IX CA commanders and men both appeared to be inexperienced and lacking in élan.[5] It was perhaps not without reason that Pétain would raise some serious doubts about the plan of attack. He wrote to d'Urbal on 13 June stating his concern that, with the plan as it then stood, XXXIII CA could very well be walking into another sharp salient, just as had happened on 9 May. If XXI CA failed to capture the *Bois de Givenchy*, or if the IX CA failed to advance far enough (and Pétain showed special reservation here, not least because of the length of ground IX CA would have to cover if it were to successfully protect XXXIII CA's flank), XXXIII CA would suffer deadly enfilade fire and lose men trying to hold an untenable

---

2 Ibid., p. 100.

3 Ibid., p. 90.

4 *SHD*, 22N573; «*Ordre Particulier à Corps d'Armée à Capitaine LAMBERT – Saint-Pol» 29 mai 1915.*

5 *SHD*, 22N573; «*Capitaine Dequa à M. le Gal Comt. la 34e Brigade» 10 juin 1915.*

position, as it had five weeks prior.[6] These concerns were legitimate, and were further strengthened by the fact that Pétain was not the only officer in Tenth Army to voice serious concern over the coming attack on 16 June.

In any case, the artillery preparation had already begun, and was being closely monitored across the whole of the front. Men in the front-line trenches along IX CA front had been issued flares with which to help register the artillery bombardment, and made frequent and detailed reports on the efficacy of the bombardment.[7] Reports during the preparation noted a large proportion of shells fired as direct hits on the major defensive features of IX CA's front, the 5 *Chemins* crossroads and the destroyed mill. These key strong-points were also targeted by artillery from 39^e^ *Division d'Infanterie*, DI (XX CA) in the hopes of assuring their destruction. While the artillery pounded away, the infantry were furiously preparing their jumping-off trenches and *boyaux*. Unfortunately, the infantry did not seem to be doing as well as the artillery was.

On 15 June General Guignabaudit, commanding 17e DI on the right of IX CA's front, wrote a letter to General Curé stating that preparations were 'not absolutely complete' and that, in any case, they did not conform to the prescriptions of Note 5779. It is worth remembering that 17e DI's preparations for its failed attack on Loos on 9 May were also not done according to Note 5779 suggesting either laziness, lack of time to prepare or an overt rejection of the official doctrine. In some places the French *bases de départ* were 200–300m away from the German front lines, as opposed to the 'less than 150m' recommended by GQG.[8] Front-line regiments had been worked to death, the note asserted, and they still had much work to do before the next day's big push. Some front-line regiments were able to be relieved on the night of 14–15 June, but would find themselves right back in the trenches and going over the top on 16 June without having had any real time to rest.

Despite the protestations and reservations of Pétain and 17e DI, an attack of this magnitude simply could not be stopped at such short notice: the artillery preparation thundered on. 114e RI (18e DI) was evacuated from its forward position so that the heavy artillery (*artillerie lourde*, AL) could focus on the German trenches immediately opposite.[9] As attritional fighting during the second phase of the battle wore on, and units desperately struggled for various key points of terrain, the opposing trench networks came exceptionally close to one another in the most fought-over areas like Neuville and Souchez. Under such circumstances incidences of friendly fire casualties escalated. This frequently lead to acrimonious verbal and written barrages between infantry and artillery commanders, with divisional or

---

6 *SHD*, 22N1832; «Le Général Pétain, commandant le 33e *C.A. à Monsieur le Général commandant la 10e Armée*» 13 *juin 1915*.

7 *SHD*, 22N573; «*Groupement Lafoud*» 13 *juin 1915*.

8 *SHD*, 22N573; «Le Général de Division GUIGNABAUDIT Commandant le 17e *Division d'Infanterie à M. le Général Com. Le 9e Corps*» *15 juin 1915*.

9 *SHD*, 22N573; «*Compte-rendu des 24 heures du 14 au 15 Juin*» 15 *juin 1915*.

corps commanders having to act as arbiter between the two.[10] The commonplace practice of retiring front-line troops to a safer position was so disruptive (not least of all because troops would often return to find their own trenches bombarded, and thus in need of time-consuming repair work) that d'Urbal, admittedly after the fact, surmised that perhaps the GQG doctrine of digging forward trenches close to the enemy front lines was not practicable in the field.

In any case, no friendly fire was suffered by 114e RI on 15 June and observers were able to report several large explosions in the 5 *Chemins* area, but also recognized that a great many German positions were now sited on the reverse slope of hills, thus making it difficult to destroy the barbed wire which lay in generous bands before them.[11] The bombardments of the 5 *Chemins*, destroyed mill, and the trenches north of Neuville all appeared to be very well sited but, in the case of the bombardment north of the village, a well-sited bombardment did not ensure an effective bombardment; reports claimed that this particular defensive network remained largely unscathed. This was to be a theme that would haunt the *poilus* of Tenth Army during the renewal of the general offensive in Artois.

The experience of IX CA in many ways mirrors the wider experiences of most Tenth Army formations: exhausted soldiers asked to take on increasing burdens in the days immediately before the attack; and artillery preparations whose results did not seem to match the intensity or accuracy of the bombardments being fired. On 13 June, the day preparatory fire began along the Tenth Army front, 360e RI (70e DI) launched an attack out from the *sucrerie* and met with mediocre results, capturing portions of the German first line at the cost of one officer wounded and 25–30 men casualties.[12] Unsurprisingly given the very close quarters at which the two forces were arrayed, the men of 360e RI fell under friendly fire for a full 30 minutes on the 13th.

Despite a largely failed attack on the 13th, three days before the general offensive was to begin, 360e RI did not stop there but attacked yet again on the 14th, capturing roughly 30–50m of trench – but severely weakening liaison with 204e RI (70e DI) in the process as the regiment continued to strive for a good place from which to launch its attack on 16 June.[13] Exhausted after three days of fighting and heavy enemy artillery bombardments, 204e RI was relieved by elements of the 13e DI (XXI CA) the night before the attack; the 360e was not so lucky. Final operation orders for the division were given in a somewhat loose manner as General Fayolle (commanding 70e DI) could not be certain exactly where his front-line troops would be when dawn rose on the morning of the 16th. This was to be the norm along many of the more active sections of Tenth Army.

---

10 *SHD*, 24N1754; «Rapport de Lt.-Colonel *Jucqueau Commandant l'Artillerie de la 70e Division au sujet d'incidents de tir survenus dans la nuit du 13 au 14 Juin, dans la matinée et dans l'après-midi du 14*».

11 *SHD*, 22N573; «*Compte-rendu des 24 heures du 14 au 15 Juin*» 15 *juin 1915*.

12 *SHD*, 24N1754; «*Compte-rendu de 14 Juin 1915 (4h45)*» *360e Regt .d'Infanterie*.

13 *SHD*, 24N1754; «*Compte-rendu des* événements *de la nuit 14 au 15 juin*» 360e *RI*.

Just as IX and XX CAs were reporting back on the well-registered fire of their artillery, XXI CA could boast the same. However, unlike IX and XX CAs, XXI CA sat proudly along the crest of Notre Dame de Lorette, and had a commanding view of the battlefield before them, thanks to the meticulous reorganization of NDL under Maistre's command following the capture of the hill.[14] Artillery–infantry liaison was generally very tight under Maistre, and his thoroughness and attention to detail would pay off in the coming offensive. Maistre actually made artillery observers a special service within his corps, for which men were individually selected and given extra training in a variety of relevant skills, such as the maintenance and repair of telephone wire.[15] Such men were to occupy the front-line trenches with the infantry, not only to observe and correct artillery fire better but also to liaise with the infantry better and act as a morale booster for the frontline *poilus*, who would have more faith in artillerymen who had to undergo the same living conditions that they did. As in other corps, the front-line infantry were given flags with which they would communicate with the artillery and give the artillery a clearer picture of where the French line ended and the German line began. Reports from forward observers, front-line infantry and advanced patrols were expected to come in detailed and often.

The system Maistre built was hard at work from 13 June in making sure the preparation was as successful as possible. Early on it had been ascertained that the German wire was 50m out in front of the German trenches, and that special bombardments were needed to destroy these emplacements.[16] Accurate reporting was made difficult as patrols were required to properly examine much of the damage done, and the Germans had undertaken a heavy bombardment against NDL, inflicting serious casualties and generally disrupting XXI CA's preparations for the attack on 16 June.[17] Nevertheless, highly detailed maps were drawn up showing the exact state of enemy defences, especially referring to the barbed-wire emplacements (Figure 6.1 below). Observations made by 43e RI found that the 75mm guns were essentially only moving the German emplacements about and were not powerful enough to destroy the *chevaux de frise* (rolls of large, pointed wooden stakes on which barbed wire was often hung) opposite them.[18] In response,

---

14 *SHD*, 24N1754; «Le Général Maistre Commandant le 21e *Corps d'Armée à M.M. le Général Cdt l'Artillerie (pour exécution). Le Général Cdt le 13e Division (à titre de renseignements)» 10 juin 1915.*

15 Ibid. «Le Général Maistre Commandant le 21e *Corps d'Armée à M. les Généraux Cdt les 13e, 43e, 48e et 58e Divisions, le Général Cdt l'Artillerie, le Lt.-Colonel Cdt le A.L. et le Lt.-Colonel Cdt le Génie*» 11 *juin 1915.*

16 *SHD*, 22N1472; «Le *Général Maistre Commandant le 21e Corps d'Armée à M.M. les Généraux Cdt les 13e, 43e, 48e, 58e et 92e Divisions, Le Général Cdt l'Artillerie, le Lt.-Colonel Cdt l'A.L.» 12 juin 1915.*

17 Ibid. «*Compte-rendu des* événements *de la Journée*» 14 *juin 1915.*

18 *SHD*, 22N1457; «Le Général Lombard Commandant la 43e *Division d'Infanterie à Monsieur le Général Commandant le 21e Corps d'Armée*» 14 *juin 1915.*

a separate bombardment by a battery of modern 155CTRs was ordered, which was effective in creating sufficient gaps in the *chevaux de frise* for the attack to have a reasonable chance of success on the 16th. The destruction of the enemy wire had been the top priority for all arms, even machine guns, under Maistre's command and everything possible was done to meet this goal.[19]

Despite the difficulties posed by the strong German defences, much improved since 9 May, mediocre observation and artillery bombardments of varied efficacy the men of Tenth Army, and especially IX CA, were going to have to give it their all. D'Urbal was intent on capturing Vimy, and ordered that the main attacking units leave no one in reserve. Instead, they were to rely entirely on army reserves to come in behind the assault waves and help consolidate any captured trenches. This was to be it, the correction of everything that went wrong on 9 May: reserves too far behind the advance columns of infantry; an insufficient mass of artillery shells; and units being cut down by deadly flanking fire from Souchez, La Folie and Neuville. In this sentiment General Cugnac, who had commanded 18e DI as it went into combat on 11 May, wrote a letter to General Curé in which Cugnac shared his experiences in the last big push with the IX CA commander:

> There's no point in saying that you will mask Souchez and Neuville – you can keep the enemy from attacking your flank – but you cannot keep machine-gun bullets from coming. You cannot mask machine guns … . The old principle of our fathers remains true – you cannot attack a wall without having destroyed the bastions' flanking organs, without demolishing the ramparts.[20]

Phase Three: The Second General Offensive

As morning came on 16 June a small group of NCOs from 125e RI (17e DI, IX CA) crept out of their trenches to inspect the damage wrought on the enemy lines by three days of solid bombardment. Looking towards 5 *Chemins* they saw the German trenches well occupied, and sitting safely behind a band of intact barbed wire.[21] Looking over towards the destroyed mill they could hardly detect any sign that a preparatory bombardment had even taken place: trenches were in good order and full of Germans; barbed-wire emplacements remained intact. These six NCOs made their report, which unequivocally stated that the artillery had generally fired

[19] *SHD*, 19N1686; «Instruction *Personnelle et Secret» 21e CA, 11 juin 1915.*

[20] *SHD*, 22N573; «Résumé de *l'attaque du 11 Mai sur la côte 140» 1 juin 1915*: «*Il ne faut pas dire qu'on masquera Souchez et Neuville – on empêchera une attaque ennemie de déboucher sur notre flanc – mais on n'empêchera pas les balles de mitrailleuses d'arriver. On ne masque pas des mitrailleuses … . Le vieux principe de nos pères reste vrai – on ne peut pas attaquer une courtine avant d'avoir détruit les organes de flanquement des bastions, avant d'avoir démo[u]lé les caponièrs.*»

[21] Ibid. «*Compte-rendu de la Journée du 16» 17e DI.*

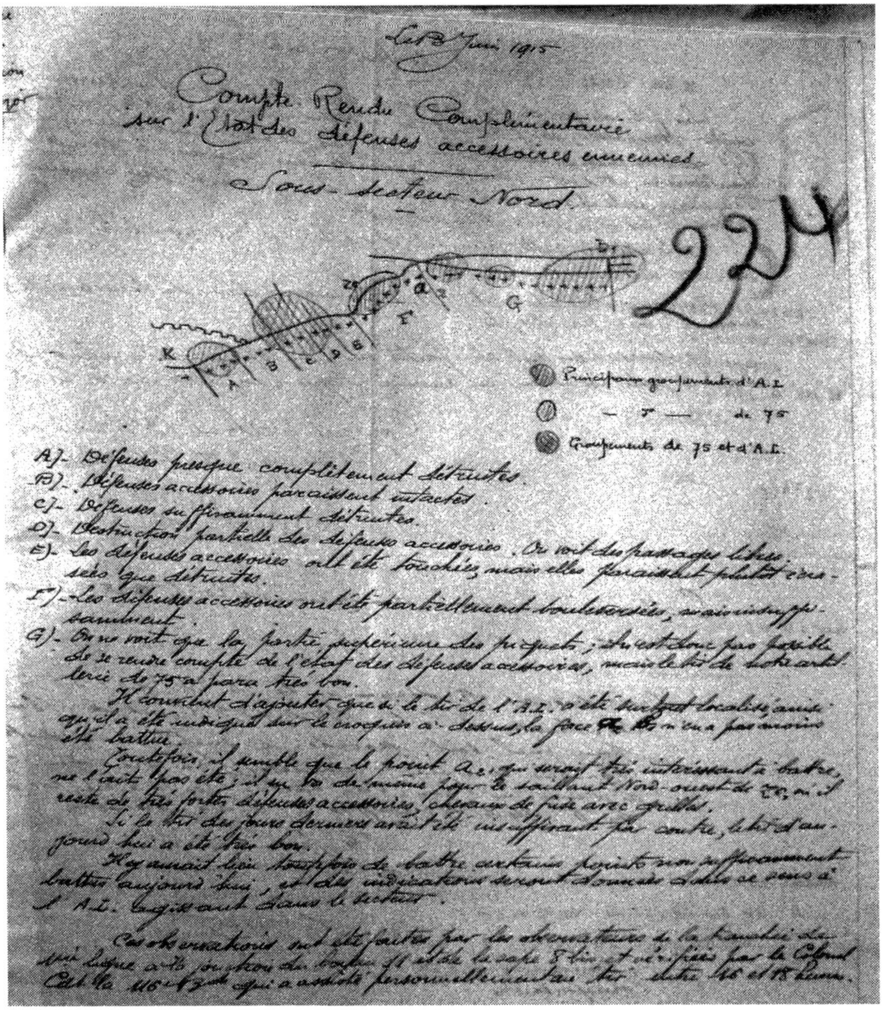
Le 13 Juin 1915

Compte Rendu Complémentaire sur l'État des défenses accessoires ennemies

Sous-Secteur Nord.

A)- Défenses presque complètement détruites.

B)- Défenses accessoires paraissant intactes.

C)- Défenses suffisamment détruites.

D)- Destruction partielle des défenses accessoires. On voit des passages libres.

F)- Les défenses accessoires ont été partiellement bouleversées, mais insuffisamment.

Figure 6.1 Example of XXI CA report from 13 June 1915

Source: *SHD*, 22N1457.

too short, thus missing the enemy wire altogether, and passed it on to the artillery, which began its final bout of preparation later that morning at 09.15.[22]

Visibility was questionable on the morning of 16 June, and the AL started its share of the preparation slowly so as not to waste ammunition. A good bombardment was reported against 5 *Chemins*, although the infantry still felt the shells were falling a little short. At 12.15 the artillery ceased its preparatory fire and

[22] Ibid.

began a hybrid barrage: one barrage rolling forward from the French front line at 50m bounds to help protect the advancing infantry, while a second barrage fired by field guns opened up at maximum range and gradually crawled back towards the French front line by 25m bounds (in the hopes of disrupting German artillery fire) until both barrages reached *cote* 140, where they would inflict a strong, sustained fire on the German defenders there before French infantry launched their assault on the ridge. At the same time the artillery began firing its barrages, *poilus* from every part of France, including Foreign Legion and colonial troops, clambered out of their forward positions and surged upon the enemy trenches.

Unfortunately, the final three hours of preparation would not prove to be any more effective than the past three days of bombardment had been, and the *poilus* of IX CA fell against largely intact, heavily defended German positions while being swept by heavy artillery and machine-gun fire. 17ᵉ DI fell back, unable to penetrate the German defences opposite them, while 18ᵉ DI at least broke in, and captured the first German position.[23] Another bombardment and a second attack were planned for that afternoon. In a surprising move for mid-1915, IX CA (as well as XX and XXXIII CAs) was given poison gas shells filled with a mixture of carbon disulphide and phosphorus, making them both asphyxiating and incendiary; their use would represent the first time that asphyxiating gas shells had ever been fired from artillery in combat. The shells were primarily concentrated against known locations of German batteries and other rear areas for the simple reason that the French had no means of protecting themselves against their own weapon; using gas against German infantry positions would have made trenches uninhabitable for defender and attacker alike.[24] This bending of tactics to logistical realities would, unintentionally, begin a growing trend which would see the use of gas increasingly reserved for the suppression of enemy guns rather than enemy infantry (which is how both the Germans and British used gas in 1915). By 1916 the use of gas in counter-battery fire would be a cornerstone of the French methodology, and gas shells would grow to represent up to 25 per cent of French shells manufactured each month.

Tenth Army had access to 10,000 gas shells on 16 June, and concentrated them against difficult urban positions like Neuville, Souchez and Angres, as well as against areas in which German artillery was known to be concentrated, like La Folie. The first use of poison gas shells in combat would prove to be highly effective. Their use against Neuville and La Folie by IX and XX CAs on 16 June would entirely suppress the German artillery opposite them for an hour and a half from 13.00 to 14.30, an incredible feat of counter-battery fire.[25] In Angres, the use of these shells produced several large fires which raged through the town, despite the shells proving to be less incendiary than had been hoped: only 10 per cent of the gas shells fired on 16 June were seen to produce a fireball. Elsewhere,

23 *SHD*, 22N573; «*Compte-rendu de l'action de la 18ᵉ DI dans la journée du 16 Juin*».

24 *SHD*, 16N707; «Note pour la D.A.» *9 juin 1915*.

25 *SHD*, 22N573; «*Compte-rendu de la Journée du 16*» *17ᵉ DI*.

especially in the shattered remnants of villages like Souchez, the shells failed to produce any incendiary effects (the buildings, by this point being little more than piles of rubble, had been rendered largely inflammable), but their asphyxiating power meant that they still played a crucial role in suppressing enemy fire. As effective as they were, however, fire-suppression only solved one problem; it could not help the infantry get through intact German wire emplacements. Despite the unprecedented suppression of the German guns as a result of the gas attack on La Folie, 17^{e} DI was able to advance only a meagre 100m, while 18^{e} DI was not able to add to their advance at all, faced with the intact second line of German trenches. This was far from the valiant success that had been expected of IX CA.

To their right, XX CA performed no better, with 39^{e} DI being entirely repulsed in their initial attack, despite the order to leave their trenches early in the hope of reaching the German line before the Germans could leave their dug-outs and establish an organized defence.[26] It had been known at least by 15 June that XX CA's artillery preparation was leaving much to be desired, and it may be that General Balfourier, commanding XX CA, hoped at least partially to offset this deficit with more aggressive infantry tactics.[27] By 15.20 39^{e} DI, having failed to profit from the temporary disruption of German artillery fire afforded by the launching of gas shells into La Folie and Neuville, was planning a new artillery bombardment in which all its artillery would be focused on the German front line with the hope of at least being able to advance that far.[28] Ultimately, even this modest goal would not be met; 39e DI would blame its inability to advance on the failure of the neighbouring 17e DI to do the same. 11e DI did not fare much better and, opposed by the same insurmountable forces of an intact enemy position and intense artillery barrage, failed to advance.[29]

To the left of IX CA, Pétain's XXXIII CA fared much better. Unlike 77e DI, the *Division Marocaine* (DM) had been given weeks to rest, recuperate, rebuild and retrain after being all but annihilated from 9 to 11 May. When the time to attack came on 16 June the men of the DM, arrayed in depth, saw before them a 'perfect' preparation, as after-action reports claimed, and easily overwhelmed the first German position, suffering only minimal casualties in the process.[30] What few Germans were to be found in their lightly held front line mostly surrendered without putting up any resistance. As the DM advanced further it became clear that the enemy of 16 June had little in common with the enemy they had fought six weeks earlier. Gone were the quantities of outdated rifles such as had been

---

26 *SHD*, 22N1358; «*Rectificatif à l'ordre Général d'opérations No. 120 pour la Journée* J».

27 Palat, *La grande guerre sur le front occidental*, p. 348.

28 *SHD*, 22N573; «*Compte-rendu de la Journée du 16*» *17e DI.*

29 *SHD*, 24N214; «Le Général de Division Balfourier Commandant la 20e *Corps d'Armée, à Monsieur le Général Commandant la 11e Division d'Infanterie*» 20 *juin 1915.*

30 *SHD*, 24N2908; «*RAPPORT du Lt.-Colonel Daugan, Commandant le 4ème Tirailleurs au sujet des opérations des 16 et 17 Juin 1915*».

captured on 9 May: the Germans were clearly all equipped with more modern, reliable weapons, and their defences, which had been fairly elementary in May, were now composed of numerous flanking positions and deep dug-outs where Germans lived in 'great comfort'.[31] Even the German sandbags were reported as being markedly better than their French counterparts': lighter, allowing a soldier to carry more than just the three a *poilu* could manage, and more durable.[32]

At the front of the DM the 4e Tirailleurs and 8e Zouaves led the charge, quickly capturing their main objective, *cote* 119, at which point enfilade fire from Souchez forced the forward troops to stop and take cover. Crucial secondary waves which were to maintain the impetus of the attack straggled behind as they advanced through *boyaux* clogged with captured prisoners and enemy wounded while being subjected to increased enemy artillery fire. By 20.00 reserves had made it to *cote* 119, and the DM had organized a defence on the hill which would fight off several violent counter-attacks over the next few days.[33] German counter-attacks were not particularly well organized, but were certainly well provisioned with grenades, which at times were landing five or six at a time in the front DM trenches on *cote* 119, inflicting heavy casualties and quickly breaking down the mental resilience of the French defenders. Despite continued pressure and an inability to communicate with their artillery, which might have seriously eased their task of defending the important high ground they had captured, the men of the DM were resolved to hold on. Little did they yet know that, just as on 9 May, they had fought themselves triumphantly into a dangerously isolated and tenuous position.

On 16 June 77e DI, along with 70e DI, had one of the most crucial tasks on which, it was thought, the success of the entire attack would rest: capturing Souchez, and thus freeing IX CA from the deadly enfilade fire that would otherwise pour out of the village. This task was made more difficult by the failure of the carbon disulphide/phosphorus shells launched by XXXIII CA to seriously affect the German positions in Souchez.[34] The resistance of the village to incendiary effects meant that few machine-gun positions could be destroyed by the new gas shells, while the asphyxiating effects were not as strong in Souchez as they had been in La Folie and Neuville. This was possibly owing to a reduced concentration as XXXIII CA split its gas shells between two main villages (Angres and Souchez); meanwhile there were two entire corps focusing on suppressing the German artillery in La Folie and Neuville. Aware of the importance of the task before them, 77e DI was eager to apply lessons the division had learned during the 9 May push. Engineers were sent forward in large numbers to help maintain the impetus of the attack and eventually help prepare the forward positions for defence, and machine

31 Ibid.

32 Ibid.

33 Ibid.

34 *SHD*, 22N1832; «*Comptes-rendus des opérations du 33e C.A. du 16 au 23 Juin 1915*». It is possible that these incendiary shells were carbon bisulphide/phosphorus shells which were both incendiary and produced poison gas upon detonation.

guns went over with the first wave to provide defensive firepower for thwarting the inevitable German counter-attacks which the division would be subjected to.[35] Speed was going to be the essence of the attack:

> For the attack to be successful every man must know that it is necessary for the front-line troops to take, in one sole bound, the successive enemy positions (in the present case, numbering two) without occupying them[.] The mopping-up of trenches ought to be done by units specially equipped with explosives, grenades, revolvers and knives, designated by corps commanders, and well acquainted with their role.[36]

The artillery bombardment was much heavier than that on 9 May, with guns allowed to fire twice as many shells in the days preceding the attack than had been available six weeks previously.[37] Unfortunately, this increase in tonnage could not counter-balance the efficacy of well-constructed trenches sited on reverse slopes. The inability of the French 75s to hit these positions was not seriously addressed until the night of 15–16 June, when the responsibility for destroying the German wire opposite 77ᵉ DI was transferred to a number of 58mm cannons.[38] These old, outdated weapons would have roughly 12 hours to do what two and a half days of field artillery fire could not.

At H-hour on 16 June the men of 77ᵉ DI would test the efficacy of their artillery bombardment with their own flesh and blood. 159ᵉ RI rose out of their trenches and stormed over the small hillock in front of them to be met, like so many other units on 16 June, by intact German wire, an intense counter-barrage and small-arms fire. By 12.30 the regiment had lost its commanding officer killed; by the end of the day every single officer from 159ᵉ RI would be killed or wounded.[39] 97ᵉ RI made much better progress, occupying the Souchez cemetery without taking many casualties in its initial assault. Unfortunately, the failure of 159ᵉ RI seriously compromised both the progress of 97ᵉ RI and the left flank of the DM, thus making any serious push on Souchez all but impossible. When General Pillot, commanding 77ᵉ DI, finally got word that 159ᵉ RI's attack had failed (news he in fact received from

---

35 *SHD*, 24N1991; «*Compte-rendu des opérations de la 77ᵉ Division dans les Journées du 16 et 24 Juin 1915» 25 juin 1915.*

36 *SHD*, 24N1991; «*Annexe à l'ordre d'attaque*» 77ᵉ *Division, 13 juin 1915*: «Pour *que l'attaque réussisse, il faut que tous les hommes sachent qu'il est nécessaire que les premiers lignes dépassent d'une seul élan les lignes successifs des tranchées ennemies (dans le cas présent, au nombre de 2) sans s'en occuper, le nettoyage des tranchées devant* être *fait par des unités spéciales munies de pétard, grenades, révolvers, couteaux, désignées par les Chefs de Corps et connaissant bien leur rôle.»*

37 Ibid. «TABLEAU *concernant l'EMPLOI de l'ARTILLERIE de CAMPAGNE du 33ème C.A. (à l'ordre no. 3859 PC du 11 juin 1915)».*

38 Ibid. 'Unnamed note'.

39 Ibid. «Colonel *Cdt la 88ᵉ de à 77ᵉ DI Compte-rendu*».

DM liaison officers) he planned a second attack for 16.00, and emphasized to his subordinates just how important it was for the 77e to make serious progress.[40] At 16.00 159e RI rose out of its trenches, and almost immediately retired back into them under heavy artillery and machine-gun fire. The serious progress Pillot wanted was simply not possible under the prevailing conditions.

With 77e DI unable to advance, the task of taking Souchez would fall to one of the workhorse divisions of Tenth Army, Fayolle's 70e DI. As has been described above, 70e DI was engaging in robust offensive action just two days before the attack, and had to borrow 237e RI from 13e DI to relieve one of the battered regiments along its left flank. As was the case across much of the rest of Tenth Army's front, 70e DI was met with a violent bombardment as it left its trenches at H-hour.[41] The 42e BCP managed to occupy part of the Château Carleul, just as it had on 9 May, despite having to cross very broken terrain and meeting stiff German resistance on the way. 42e BCP was unfortunately required to stop its advance short so as to maintain liaison with 97e RI (77e DI). 360e and 237e RIs did not fare quite so well, with only one company of 237e making any progress at all against the impenetrable wall of fire they were confronted with.[42] The achievement of this one company is somewhat dimmed when one considers that it was heavily profiting from its proximity to the more successful 13e DI. In all, this should not have been surprising considering the intensive action 360e RI had seen in the days leading up to 16 June, and the fact that 237e RI had been in the front-line trenches for only a day before the attack was launched.

Further to the left is found one of the few genuinely positive sectors of Tenth Army on 16 June. With a commanding view of the German defences opposite them, experienced men and attentive officers the XXI CA was in a good position to advance on its objective, the *Bois de Givenchy*. At 12.15 43e DI blew the mine it had dug under the German position opposite and, advancing quickly, occupied the crater and began preparing it for defence against any counter-attack without suffering undue losses.[43] To its right, 48e DI made the best progress of 16 June after the DM, advancing nearly 1 kilometre and taking all of its primary objectives within only 25 minutes.[44] Such success did not come free, however, and losses were heavy. To their right, 13e DI – moving against the same difficult German defensive works protecting Souchez as 70e DI – had advanced 150m before German artillery and machine-gun fire stopped the attack cold. Casualties mounted, and by afternoon the two companies which led the 13e DI attack were being led by a single *sous-lieutenant*.[45]

40 *SHD*, 24N1991; «*77e DI à 88 Bde*».

41 *SHD*, 24N1754; «*Rapport à la suite des Opérations du 16 Juin*» 70*e DI.*

42 Ibid.

43 *SHD*, 22N1457; «*Compte-rendu de fin de journée*» 16 *juin 1915.*

44 *SHD*, 22N1457; «*Compte-rendu des* événements *de la journée du 16 Juin*» 48*e DI.*

45 *SHD*, 22N1457; «*Compte-rendu de fin de Journée*» 13*e DI, 16 juin 1915.*

17 June would prove to be a somewhat different day than its analogue, 10 May. As few units had made serious progress, the business of organizing a new artillery bombardment and attack would not be as problematic in terms of communication; the loss of so many men and officers, however, would seem to negate any advantage this conferred to the French. The attack would be general, but with special emphasis on 77ᵉ DI and IX CA, both of which needed to make significant progress if the attack was to have the slightest chance of capturing Vimy Ridge. 159ᵉ RI was in a very bad state by the 17th, having suffered horrible casualties in its abortive attacks the day before.[46] Likewise, intense German artillery fire had caused serious casualties to 97ᵉ RI and the DM, which was beginning to consider pulling back into a more secure position.[47]

The attack was slated for 16.00, but owing to the difficulty of providing an adequate artillery bombardment the attack was delayed until 16.30. As intense German artillery bombardment made communications slow and difficult, only the 16th Company of 159ᵉ RI received word of the delay (they had lost their commanding officer and a large portion of their men to German artillery just before the attack, and, in their confusion, failed to attack).[48] The rest of the regiment went over the top, and met with intact German trenches and ready defenders. As the infantry attack stalled, the renewed French preparatory bombardment kicked in, forcing the men of 159ᵉ RI to take whatever cover they could find in no man's land, trapped by German and French artillery fire.[49] Surprisingly, the 16th Company did in fact go over the top at 16.30 and, relying heavily on the use of grenades, managed to take and hold the first line of German trenches. A night attack was later planned to bring 159ᵉ RI forward into better liaison with the *chasseurs* on their right. The troops advanced silently into no man's land until they were within 30m of the German trenches. When they popped up and launched their surprise attack they were met with still intact German wire, and suffered heavy losses ultimately capturing a mere 60m of the enemy trench network.[50]

This story is the blueprint for most of the action seen on 17 June, with 70ᵉ DI and XXI CA also making some progress (70ᵉ DI actually captured two successive lines of German trenches on the 17th), but at heavy cost in the first attack and utter failure in the second. Another common story was that of German preparedness. IX CA bombardment lasted from 15.00 to 16.00. At 16.00, as the men of IX CA left their trenches to attack the German front line, they saw a series of flares go up in front of them, and were almost immediately hit with a dense German artillery bombardment which was closely followed by German machine-gun chatter.[51]

---

46 *SHD*, 22N1832; «*Compte-rendu des opérations du 33ᵉ C.A. du 16 au 23 JUIN 1915*» *2 juillet 1915*.

47 *SHD*, 24N1991; «*Télégramme: 88ᵉ Brigade à 77ᵉ Division*» *17 juin 1915; 19.00.*

48 *SHD*, 26N701/1; 159ᵉ RI JMO.

49 Ibid.

50 Ibid.

51 *SHD*, 22N573; «*Compte-rendu de la Journée*» 17 *juin 1915*.

French gas shells had been launched against La Folie but, as happened on the day before, the use of gas against German batteries could do little to support advancing French infantry as they encountered ready German infantry defending intact positions. Neither IX nor XX CA managed to make any appreciable advance.

By 18 June it was obvious that things were going horribly wrong. D'Urbal decided to subordinate everything to the main thrust on Vimy Ridge, which he still hoped to capture, despite the inability of IX CA to make even the most meagre of advances. IX CA was ordered to not waste any men on Neuville, but instead to focus all its attention on Vimy Ridge (an order he neglected to make XX CA aware of). When General Balfourier of XX CA found out that IX CA would not be linking up with the efforts of XX CA he flatly refused to attack; Balfourier would not suffer his already battered troops to advance without their left flank secure. The forward troops of the DM had pulled back during the night and the battle-worn 4e Tirailleurs had been relieved.[52] While this put the DM in a much safer position, it also meant that on 18 June it would effectively have to advance into terrain it had already deemed as being too dangerously forward and isolated.

The story of 18 June is a very simple one that does not need much exposition, in light of the previous two days. 18e DI reported what was the experience of most units along the front, saying that they could not effectively bombard the German positions as they were sited on reverse slopes.[53] Terrestrial observation was poor and troops essentially had to be sent against positions whose status was largely unknown, which led to a great number of units running into intact wire and ready defenders. In such a situation heavy casualties were suffered and no advance was made.[54]

After the 18 June attack Foch called off the offensive, an order that would have to be repeated by Joffre a week later before the Artois sector would fully calm down.[55] Tenth Army had lost somewhere in the region of 100,000 to 121,000 casualties since 9 May while inflicting anywhere from 50,000 to 80,000 German casualties in the process.[56] A variety of explanations for the failure of Tenth Army to capture Vimy Ridge despite having what, at the time, seemed like overwhelming amounts of quality troops, guns and munitions cropped up almost before the battle ended. Joffre felt that Tenth Army would have been perfectly successful if only it had had more artillery; if reserves had been able to move forward more quickly; and if the British army had been more successful in the early phases of the battle, especially

---

52 *SHD*, 24N2908; «*RAPPORT du Lt.-Colonel Daugan, Commandant le 4ème Tirailleurs au sujet des opérations des 16 et 17 Juin 1915*».

53 *SHD*, 22N573; «Le *Général Lestoquoi Commandant le 36e Brigade d'Infie à M. le Général Comt la 18e DI» 18 juin 1915*.

54 Ibid.

55 Doughty, *Pyrrhic Victory*, p. 164.

56 *AFGG*, *Tome III*, p. 100; Daille, *Histoire de la Guerre Mondiale*, p. 123; Doughty, *Pyrrhic Victory*, p. 165; and Palat, *La grande guerre sur le front occidental*, p. 365.

on 9 May.[57] Foch partially felt that the German artillery domination had stopped the French from marching victorious on Vimy (at least in the June attacks, as Foch claimed that from 20 May onward the Germans had clear artillery superiority in the theatre). More importantly, Foch expressed serious doubts that fighting Second Artois as a 'breakthrough battle' had left any chance for a serious success to be had, even if he also claimed, retrospectively, that Second Artois had as its goal to keep reserves from heading to the Eastern Front, and that the battle succeeded in that purpose.[58] Pétain was simple, putting it all down to the 'deplorable methods' used in much of Tenth Army, and Fayolle, as usual, was scathing:[59]

> Foch and d'Urbal did not take any account of the state of the battlefield or the necessities of trench warfare. They still figured it was like attacking in open terrain, and did not worry themselves about a long and detailed preparation, which is indispensable.[60]

Despite five weeks of jostling for position and substantial reinforcements, Tenth Army could not make any real progress from 16 to 18 June. Perhaps blinded by the success of 9 May, the French did not make any serious alterations to their tactics in preparation for 16 June; even if they had, the tempo of operations was such that units on the ground would not have had sufficient time to train in them before going over the top. Some of the doctrine which came down from GQG in mid-June certainly did not imply that great tactical strides had been made past the initial reflections after 9 May. GQG wrote on the use of the 58mm gun, stating that owing to its slow rate of fire and inaccuracy it must be used in very high concentrations in order 'to assure moral and material results of the first order'.[61] GQG was also quick to restate the fact that infantry attacks against established trench networks could in fact be successful, provided there had been an effective preparation in terms of artillery bombardment, general logistics and terrain management.

Less stellar is the annex written for Note 5779 on 18 June, which states emphatically that 'the arm of exploitation is the cavalry; its use is therefore envisaged … as [it would be] in mobile warfare'.[62] The aims of cavalry are listed

---

57 *AFGG, Tome III*, p. 103; *AFGG, Tome II*, Annexe, p. 503; and Daille, *Histoire de la Guerre Mondiale*, p. 127.

58 Daille, *Histoire de la Guerre Mondiale*, p. 127.

59 Goya, *Le processus d'évolution tactique de l'armée française de 1871 à 1918*, p. 279.

60 Fayolle, *Cahiers secrets de la Grande Guerre*, p. 113: «*Ils se figurent encore qu'on attaque comme un terrain libre et ne se doutent pas de la préparation longue et minutieuse qui est indispensable.*»

61 *SHD*, 16N1677; «Note pour les *Armées*» 10 *juin 1915; no. 2148*: «*Leur emploi en masse dans le préparation des attaques assure des résultats moraux et matériels de premier ordre*».

62 Ibid. «Annexe de la Note *du 16 Avril 1915 sur le But et Conditions d'une action Offensive d'Ensemble*» 18 *juin 1915; no. 3192*: «*l'arme d'exploitation rapide est la*

as disrupting enemy artillery and capturing enemy generals, as if the trenches stretching from Switzerland to the Channel simply did not exist. Despite archaic proclamations such as this there is much to be said about the doctrine under which the men of Tenth Army operated from May to June 1915. The doctrine might not have been perfect (most ideas were not yet fully formed in 1915, and many distinctly poor ideas were still in circulation), but enough good information was disseminated down to the rank and file for the effective skills to be given a chance to stick and become standard practice, while ineffective procedures would be dropped altogether. This could be a great strength or a great weakness of French doctrine, depending on one's opinion. Intelligent commanders would no doubt have taken the best ideas, merged them with their own and created a viable system, in the style of Maistre or Pétain. On the other hand, less intelligent commanders would be faced with the onerous task of meeting unrealistic standards, and would be prone to making grave errors of judgement in doing so too strictly. While the French struggled to evolve their admittedly forward-thinking doctrine from April 1915, the German defensive tactics were refined at an accelerated pace, eventually resulting in a system which the French were not in any way prepared to deal with. The result? The result was none other than the massacre of French troops who relied on two-month-old offensive methods against up-to-the-minute German defensive preparations and procedures. In warfare which evolves as quickly and constantly as did the fighting in the First World War, methods that are two months old could be hopelessly out of date and cost the lives of thousands of men.

In addition to the doctrinal growing pains from which the French army suffered in 1915, they also had difficulties dealing with their allies. Given the cooperation of motivated individuals, the French and British could cooperate very well in the field. In the absence of such individuals, or when inter-allied friction rose to boiling point, the system would quickly break down and become a thin veneer of unity and cooperation covering a rotting structure underneath. Both the problems with inter-allied relations and the formation of an effective and agreed-upon doctrine were in essence the same problem: the lack of form. The French army had yet to adopt a definite doctrine which could be easily communicated to all ranks, and which contained effective methodology that was not in need of much alteration by commanders in the field. Likewise, the lack of official machinery for inter-allied liaison meant that any and all liaison was *ad hoc*, and thus subject to the whim of a large number of individuals interacting without concert. The theoretical basis for winning the war on the Western Front, for beating the trenches, and doing so with all the allies acting with one mind and one goal, was there in 1915; but it would take time before these loose and scattered ideas would become the hard reality which eventually led to Allied victory.

---

cavalerie; son emploi doit donc être *envisagé et* étudié *sur le front d'attaque ... comme dans la guerre de campagne*».

Chapter 7
A Tale of Two Divisions, Part II

The Attack on Chantecler

The days immediately preceding the general offensive on 16 June were relatively calm along 34e *Division d'Infanterie* (DI's) front, with the Germans inflicting only their 'habitual' bombardment through which the division carried on improving its trench system.[1] Several false attacks were launched in the days leading up to 16 June in an attempt to confuse the enemy as to the exact time and date of the real assault. Typically, the German response was only a moderate barrage fired to block the advancing infantry.[2] The hope, of course, was not only to deplete the German stores of ammunition, but also to encourage the German artillery to distrust the German infantry and perhaps hesitate to intervene when the actual French attack emerged. Meanwhile, the French artillery maintained a steady preparatory bombardment on Chantecler, as it had since early June. By 14 June the artillery had done significant damage to the salient protecting Chantecler, but had as yet failed to make many clean breaks through the dense enemy wire.[3] Doing so was made increasingly difficult by the ability of the Germans to do at least a modicum of repair work at night, which greatly hindered the French artillery preparation even as their efforts increased in the days immediately before the attack.

At 12.15 on 16 June 34e DI attacked the German positions in and around Chantecler along a narrow 1,200m front.[4] The attack was launched by surprise, with the French artillery only opening up one minute before the infantry went over the top. As expected, the Germans unleashed a barrage upon the French front trenches which succeeded in seriously disrupting the attack. While one of the two lead battalions – 1er Battalion (Btn), Leixelard of the 83e *régiment d'infanterie* (RI), 67e Brigade (Bde) – managed to advance, the other (3e Battalion, Labourdette) was pinned down in its own trenches, utterly immobilized by German fire.[5] 3e Battalion, of 14e RI, advanced hand in hand with Btn Leixelard and found itself in the German front-line trenches.[6] Immediate German counter-attacks materialized, but were beaten back by French artillery. As the day wore on, however, the battalions

1 *SHD*, 26N326/2; 34e DI JMO.

2 Ibid.

3 Ibid.

4 Ibid.

5 *SHD*, 24N741; *«Compte-rendu de 15h7» 67e Bde, 16 juin 1915.*

6 *SHD*, 24N741; *«Compte-rendu» 67e Bde, 16 juin 1915; 16.20.*

began to run out of essential trench-fighting tools, such as hand grenades, and were eventually forced to pull back after a mere four hours in the enemy trenches.[7]

To the left of 83e RI (67e Bde), 88e RI (68e Bde) did not fare well. Despite being met with the same thick German counter-bombardment, the right of 88e RI made it to the German wire, which they found severely damaged yet still impossible to cross.[8] The left of the 88e failed to leave their trenches for more than a moment before German artillery obliged them to return to the relative safety of their *parallèles de départ*. Further attempts were made at 17.30 and 19.15. Both were stopped cold by German machine-gun and artillery fire, although the 19.15 attack could boast of having advanced the French infantry a princely 50m into no man's land, with some units going so far as to capture a few German saps. 68e Brigade would go on to attempt a pair of night attacks which were repulsed by enemy small-arms fire.[9] Despite a serious and concerted effort over several weeks 34e DI had not managed to produce results any better than they had earned on 9 May.

While the attack had undeniably failed it was, nevertheless, substantially less costly than the failed attack on 9 May. Whereas 88e RI had suffered 1,535 casualties on 9 May it suffered a comparatively light 333 casualties on 16 June: a result, primarily, of the sheer unwillingness of its men to advance into the German barrage.[10] 67e Brigade faired considerably worse, losing some 944 casualties, with Btn Leixelard being all but annihilated in its unsuccessful struggle to maintain its forward position against repeated German counter-attacks. As on 10 May, 34e DI was paralysed in confusion on 17 June. Units were mixed and disoriented, while the trenches and *boyaux* were cluttered with bodies, living and dead, which made communication and transportation especially difficult.[11] Nevertheless, de Lobit planned an attack on the 17th whose complexity barely surpassed the 'same troops – same objective' attack orders of a month earlier.[12] Again the attack was to be launched 'by surprise', with no preceding artillery bombardment, and would involve the same units attacking the same objectives as they had the day before. This plan of attack was roundly vetoed by 67e Bde, who successfully argued that they were in absolutely no position to organize a new attack.[13] To make matters worse the Germans had passed a relatively calm evening, and had repaired many of the breaches through which 34e DI had advanced the day before, thus forcing the AD/34 (*artillerie de division*) to essentially start from scratch.

---

7 *SHD*, 26N326/2; 34e DI JMO.

8 *SHD*, 24N741; *«Compte-rendu des* événements *de la journée 16 Juin»* 68e *Brigade, 17 juin 1915.*

9 Ibid. «Compte-rendu» 34e *DI, 17 juin 1915; 04.30.*

10 Ibid. *«Compte-rendu des* événements *de la journée 16 Juin»* 68e *Brigade, 17 juin 1915.*

11 *SHD*, 24N741; «Compte-rendu» 34e *DI, 17 juin 1915; 04.30.*

12 Ibid. *«Ordre général d'opérations No. 4/8» 34e DI, 17 juin 1915.*

13 Ibid. «67e *Bde à 34e Division» 17 juin 1915; 07.00.*

The men of 34e DI spent 17 June under a constant hail of German artillery fire, which prevented any serious work being done. Over the course of the day 88e and 83e RIs were relieved by 14e and 59e RIs, respectively, who would continue the attack on 18 June. As they came into the line they were met with an unsettling event that occurred early on the morning of 18 June: a heavy German shell struck a 34e DI ammunition depot at St Nicolas, blowing up 130,000 rifle rounds and an unknown quantity of hand grenades.[14] Ultimately, the formidable state of the German defences and the inability of the AD/34 to make any significant impact forced the cancellation of the 18 June attack, thus sparing the men of the 14e and 59e RIs and ending 34e DI's long and difficult experience during Second Artois.

In Memory of Barbot: 77e DI's attack on Souchez

On 1 June 77e DI returned to the trenches. Unfortunately, the situation on the division's front had not drastically improved since the days immediately following 11 May. General Pillot wrote angrily to Pétain bemoaning his situation from the first day the principal units of 77e DI (97e and 159e RIs) reoccupied their forward trenches. In order to make room for IX CA, which would be making the main assault on Vimy Ridge on 16 June, 77e had been shoe-horned into a very tight space, sharing a mere 1,200m with the Moroccan Division (DM) – leaving roughly 600m per division, only half what they had had on 9 May. As a result, men from the two divisions became hopelessly intermingled, and both were faced with enormous logistical difficulties.[15] There was no serviceable road connecting the rearmost trenches and the supply network of XXXIII CA, which meant that supplies had to be moved by cart or by hand over broken and confusing terrain. Furthermore, there were no *boyaux* leading directly from the front to 77e DI's rear area, a direct result of the division being squeezed into a narrow space between its own former sector and that of 70e DI.[16] The only viable route took a circuitous detour south through Carency, which made the already long and perilous journey to and from the front significantly longer. These transportation difficulties did not pose merely logistical problems but tactical problems as well, making it very difficult, if not impossible, for reserve troops to intervene in the event of a German attack.[17] To help alleviate the situation Pillot strongly asked Pétain for permission to extend his front northwards, which was eventually granted.

Pillot set to work immediately rebuilding his cramped sector. Starting on 2 June he required units resting in the rear to provide work-parties to help with the construction of new trenches: 500 men each from two battalions of 67e Bde were

---

14 *SHD*, 26N326/2; 34e DI JMO.

15 *SHD*, 24N1991; «Le Général Pillot, Commandant la 77e *Division à Monsieur le Général Commandant le 33e Corps d'Armée*» 1er *juin 1915.*

16 Ibid.

17 Ibid.

ordered to work on the night of 2–3 June, a number which increased to 1,500 in total on the following night. Such a robust, if not downright exhausting, work-plan posed its own logistical problems. The relief of the *bataillons de chasseurs* à pied on the night of 3–4 June took several hours as the *chasseurs*, and the troops relieving them, had to navigate congested *boyaux* filled beyond capacity with territorials and others working on the trenches, while resupply teams were moving materiel forward and engineers were darting about overseeing the whole affair.[18] The *chasseurs* were not fully relieved until 02.00 on 4 June, and owing to the desolation behind the front lines they had no sufficient place for cantonments. Command posts were small and hard to come by. There was no cover even for the ambulances serving the division, and the *chasseurs*' field kitchen had to be moved into the rearmost trenches as there was no other place safe from German bombardment where it could be located; finding housing for relieved troops was impossible.[19]

Similar to 34e DI, the 77e also faced the problem of men becoming hardened, or at any rate, indifferent to wholly unsanitary and dangerous conditions. On 7 June Pétain wrote furiously to ask why corpses were being left out in the open to rot, decrying what he saw as an 'incomprehensible lack of respect for a soldier fallen for *la Patrie*'.[20] Pétain insisted that the bodies be carried to the rear and given the burial they deserved. The next day a slightly more pragmatic note was distributed by Pillot in which he talked about the intense bombardment churning up the buried dead, thus releasing strong odours and creating an unsanitary situation.[21] Pillot recommended a much wider use of disinfectant, but also followed Pétain's lead in suggesting the bodies be carried to the rear 'as if they were wounded', and buried at a site to be chosen by the divisional doctor.

Throughout these difficulties Pillot exercised exceptional, if stern, leadership. Much like Barbot before him, Pillot took a deep interest in the evolution of the trenches in his sector, and very closely tracked their progress in the hope of maximizing the amount of work the division could accomplish before any renewed attack. This included reprimanding soldiers for digging 'illegal' individual shelters, which he felt took time away from the construction of larger, unit-sized dug-outs that offered more protection from enemy bombardments.[22]

---

18 *SHD*, 24N1991; «Le Lieutenant-colonel Laignelot, Commandant le Groupe des Bataillons de Chasseurs à Monsieur le Général Commandant la 77e *Division; compte-rendu*» 3 *juin 1915; 05.00*.

19 *SHD*, 24N1991; «Le Lieutenant-colonel Laignelot Commandant la Groupe des Bataillons de Chasseurs à Monsieur le Général Ct. la 77e *Div.» 4 juin 1915*.

20 Ibid. «Le Général Pétain, commandant la 33e *C.A. à M.M. le Général commandant la 70e Division, 77e Division, Division Marocaine*» 7 *juin 1915*: «*un manque de respect incompréhensible à l'égard d'un soldat tombé pour la Patrie*».

21 Ibid. «77e *Division à 88e Brigade, Chasseurs, A.D/77, Génie, Btn-dépôt, Santé, Cdt du Champ de Bataille*» 8 *juin 1915*.

22 Ibid. «77e *Division à 88e Brigade, Chasseurs, Cdt du Cham de Bataille*» 6 *juin 1915*.

Pillot also closely commanded his engineers to ensure that the trenches allowed for as rapid a circulation of men and materiel as possible. Precise orders were given concerning the naming and sign-posting of various trenches and *boyaux*, as well as ensuring that trenches and *boyaux* were dug to the desired size and shape.[23] Flanking positions similar to the Germans' were dug under Pillot's direction, and front-line trenches were improved to allow for all of the front-line troops to fire simultaneously – an act which had in many places been impossible up to that time.

On 13 June artillery preparation for the 16 June attack began in earnest. That same day General Pillot wrote to his subordinates to emphasize a few important tactical prescriptions. Above all else he emphasized the need for the men of 77^e^ DI to advance beyond the German first position in and around Souchez (the second was on Vimy Ridge and in Givenchy) and certain notorious strongholds (primarily Souchez cemetery) in the first phase of the attack.[24] Undoubtedly based on the experiences the division had had in trying to capture Souchez and the Château Carleul, Pillot recognized that if the foundational strong-points of the German defence network were not overrun in the opening phase of the attack they would be reinforced and made exponentially more difficult to take. Assault waves were exhorted to advance as fast and far as possible and leave *nettoyage* to the *nettoyeurs*, who Pillot described as 'special units supplied with petards, grenades, revolvers, knives, designated by the corps commanders and knowing well their role'.[25] Machine guns, sandbags and barbed wire were all to be taken over the top by the first wave to ensure that they had the necessary materials on hand to immediately begin to consolidate their forward positions when the time came. Pillot went on to order that troops be given a variety of very specific designations outside the normal realm of tactics, making some troops responsible for controlling captured prisoners, others for directing traffic once the battle got underway and so forth.

The artillery preparation leading up to 16 June was organized with customary precision and attention to detail. Owing to the poor terrestrial observation, exacerbated by constant German artillery bombardment that made scouting all but impossible, XXXIII CA had to rely heavily on aerial reconnaissance for artillery registration and fire correction.[26] 77^e^ DI was in a particularly difficult position in terms of observation owing to being situated immediately opposite Souchez, whose remaining buildings and rubble heaps posed a serious obstacle not only for artillery observers, but also for advancing infantry who would have to clear Souchez 'house by house'. Incendiary/asphyxiating gas shells (filled with carbon disulphide/phosphorus gas) were launched into Souchez during the preparatory bombardment in the hope of igniting the village and thus driving the German

23 Ibid. «*77e Division à Cdt du Génie Divisionnaire*» 10 *juin 1915*.

24 *SHD*, 24N1991; «Annexe à *l'ordre d'attaque*» *77e Division; 13 juin 1915*.

25 Ibid. *«unités spéciales munies de pétards, grenades, revolvers, couteaux, désignées par les chefs du Corps et connaissant bien leur rôle»*.

26 *SHD*, 22N1832; «*Comptes-rendus des opérations du 33e C.A. du 16 au 23 Juin 1915*».

defenders and their well-sited machine guns from their convenient strongholds; unfortunately, the village failed to burn.[27]

The preparation was further hindered by the necessity of evacuating forward trenches before a bombardment could take place in order to avoid friendly fire, which exhausted front-line troops and often delayed or complicated artillery bombardments. Most pressing was the inability of 77ᵉ DI artillery to accurately bombard the German position opposite 159ᵉ RI, as it was sited on a reverse slope. The complete failure of the preparatory bombardment was only discovered on the night of 15–16 June, at which point two 58mm cannons were rushed up to try to break through the wire: the infantry would not know how successful this last-minute effort had been until they went over the top the next day.[28] To make matters worse, the German artillery had been very effective in inflicting casualties on the French infantry in the days leading up to 16 June, with some units losing 10 per cent or more of their combat force before the attack was even launched.[29]

At 12.15 on 16 June Tenth Army made its second attempt to capture Vimy Ridge. On 77ᵉ DI's right the lead battalion of 159ᵉ RI immediately ran into trouble. Their attack was met with intense small-arms fire, while the German heavy artillery fired a barrage behind the advancing French infantry, trapping them between two impassable walls of fire.[30] The trenches opposite the 159ᵉ were largely intact, which afforded the German defenders substantial protection from the French assault. By 12.30 the regiment had lost its commanding officer, and by the end of the day every single commissioned officer in the 159ᵉ was either dead or wounded.[31] Casualties were very high, and many reserves could not make it forward to assist the assault troops owing to the strong German counter-barrage. Nevertheless, a few companies were able to advance and take the first German trenches, just barely maintaining liaison with elements of the Moroccan Division on their right.

Initially 97ᵉ RI fared much better, taking the Souchez cemetery in their opening attack and suffering only minimal casualties. There it consolidated the captured German trenches and began preparation for a second attack against the *ilôt* south of Souchez (thus maintaining contact with 42ᵉ BCP attacking the Château Carleul) and Souchez itself.[32] Unfortunately, the poor performance of both 159ᵉ RI to its right and 70ᵉ DI further to its left brought 97ᵉ RI under increasingly concentrated German artillery fire. This made organization of a second attack difficult, and whittled away the offensive strength of the regiment . As reports began to reach Pillot the situation became clear. The failure of 159ᵉ RI had placed both the DM and 97ᵉ RI in highly untenable positions, thus threatening the progress made that

---

27 *SHD*, 22N1832; «*Comptes-rendus des opérations du 33ᵉ C.A. du 16 au 23 Juin 1915*».

28 *SHD*, 24N1991, untitled document.

29 *SHD*, 26N701/1; 77ᵉ DI JMO.

30 Ibid.

31 *SHD*, 24N1991; «Colonel *Cdt la 88ᵉ de à 77ᵉ DI compte-rendu*».

32 *SHD*, 24N1991; «77ᵉ *Div. à 139ᵉ Brigade*» *16 juin 1915; 17.00.*

day and impeding any new assaults by the more successful units. In response, Pillot organized a renewed attack to be launched in the afternoon; this attack would ultimately fail.[33]

Despite having suffered heavy losses on the 16th, especially in 159e RI, 77e DI would be obliged to continue the attack on the 17th in the hope of better supporting the Moroccan Division and finally capturing Souchez. As early as the evening of 16 June trench guns, primarily 58mm cannons, were being moved in behind the newly captured position of 97e RI, where they would be able not only to better support 97e RI (trench guns had a very short range and needed to stay close to the front lines), but would also be able to offer enfilade fire against the German positions opposite the 159e.[34] Initial plans for a renewed attack surfaced at 00.45 on 17 June, with the attack scheduled for 03.30;[35] 45 minutes later this plan was scrapped. Advances made by 70e DI and XXI CA altered the situation on 77e DI's left, thus allowing a stronger emphasis on operations to better link up with the Moroccan Division. Pillot ordered the front to be extended eastwards to this end, a trying task which was only completed at 07.00 on the 17th, and instructed Colonel Wurster (commanding 88e Bde) to work up a new attack plan and submit it to divisional headquarters by 08.00 that day.[36]

Ultimately the attack on the 17th would be planned for 16.00. The artillery preparation was difficult as the situation was essentially unchanged and mortars would have to do a large share of the work in order to effectively deal with the wire opposite the 159e. This, unfortunately, was not without its problems. The guns of 97e RI not only had to be relocated forward, but were also required to re-register against the various German positions, tasks which were both demanding and time-consuming for the gun crews. The 37mm cannons had particular trouble ascertaining the exact conditions of the front, which precluded their being moved forward. Firing was impossible until they could receive good information as to the disposition and location of the forward French troops.[37] The difficulty in organizing a reliable artillery preparation forced Pillot to delay the attack by half an hour (from 16.00 to 16.30) to ensure an accurate and effective bombardment of the German trenches. Owing to the very poor state of 77e DI's rear areas the order was delayed in getting to the front. Some elements of 159e RI did not receive the rectified order until 16.05, which resulted in some French troops going over the top at 16.00 completely isolated and subjected to both French and German artillery shells.[38]

---

33 Ibid. «*77e Div. A 88e Bde*» 16 *juin 1915; 16.00.*

34 Ibid. «Général Ct. 77e *DI à Lt. Ct. L'Ae du 58» 16 juin 1916; [expédié] 17.20.*

35 Ibid. «*77e Division à 88e Brigade, AD/77, Art. de tranchée, Génie*» 17 *juin 1915; 00.45.*

36 *SHD*, 24N1991; «*77e Division à 88e Brigade, Artillerie, Art. de tranchée, art. de pet. Calibre, Génie*» 17 *juin 1915; 01.30.*

37 Ibid. «*Artillerie Légère 77 DI (Ct. De 37) à M. le Général Commt la 77e DI» 17 juin 1915; 11.00.*

38 *SHD*, 26N701/1; 159e RI JMO.

When the division properly attacked at 16.30 the results were little different from those on the 16th. Despite a strong German artillery response and enfilade machine-gun fire pouring in, 159e RI (including 14th Company, which had been stranded in no man's land since 16.00) managed to advance up to the German wire, which was found to still be intact.[39] 97e RI fared no better against a thick German counter-barrage and violent flanking fire, and could not make any appreciable progress. The most successful elements of each regiment advanced no more than 100m, and much of this had to do with the tenacity of their grenadiers more than anything else. Even these gains were slowly whittled away as the grenadiers ran out of grenades, and the Germans began to slowly push back the exhausted French infantry. This meagre result came at a resoundingly heavy toll: 97e RI was completely exhausted, and casualties continued to mount in the face of a steady German bombardment. 159e RI was horribly disorganized and found itself being led by a captain still awaiting his temporary battlefield commission, the highest ranking officer left in the regiment.[40]

Nevertheless, the division was ordered to spend the night of 17–18 June clearing cluttered *boyaux*, re-establishing communications between the front and the rear, and ensuring better lateral liaison.[41] Another pair of 58mm cannons was ordered forward into the old front lines behind 97e RI, and the division was expected to be ready for another push on the 18th.[42] 97e RI was to continue its drive into Souchez with an attack on trench Y, while 159e RI was, à tout prix, to retain liaison with the DM by attacking the eastern edge of the village. In the rear the divisional headquarters had been moved forward to Caucourt, where there was found not a single intact structure capable of housing the entire staff.[43] While this relocation proved disruptive and uncomfortable, it gave the division access to the roads and crossroads in 70e DI's sector that 77e DI had so badly needed.

During the night of 17–18 June 159e RI, along with 61e BCP, launched an attack in the hope of maintaining their threadbare liaison with the Moroccans. The assault companies attacked by surprise and managed to advance to within 30m of the German trenches without being seen.[44] From there they made a bayonet charge headlong into intact German wire emplacements. After some confusion a small gap in the wire was located and the troops surged through, clearing their way with grenades and eventually advancing 90m into the German trench network.[45] At this point the now-familiar story began to unfold. The assault companies were subjected to an intense German bombardment and repeated German counter-

39 Ibid.

40 *SHD*, 24N1991; «*77e Division à 33e Corps d'Armée, compte-rendu*» 17 *juin 1915.*

41 *SHD*, 24N1991; « *77e Division à 88e Brigade, Artillerie, Art. de tranchée, Art. légère, Génie*» 17 *juin 1915; 19.45.*

42 Ibid.

43 Ibid. «Note de Service» 17 *juin 1915; 19.00.*

44 *SHD*, 26N701/1; 159e DI JMO.

45 Ibid.

attacks. As the men ran out of grenades and casualties mounted they began to slowly give up ground, eventually ceding 30 of their 90 captured metres.[46] For their heavy losses the assault companies earned a toe-hold in trench D, which the regiment had been attacking since 16 June, thus marginally improving the regiment's odds of success on the 18th.

77e DI's attack on 18 June was made under ostensibly the same conditions as the previous day's, and met with similar results. As the men of 77e DI were preparing for their 17.00 attack a German reconnaissance plane buzzed overhead, undoubtedly watching the great movement of troops towards the front.[47] When the *poilus* left their trenches they were greeted with a violent German counter-barrage and small-arms fire, which inflicted heavy casualties on the advancing troops. The men doggedly pressed on until they reached the German wire, which remained largely intact. There they resorted to very heavy grenade use as the men filtered through narrow gaps in the enemy wire or entered the enemy trenches via German saps.[48] Their advance was minimal, and their casualties painfully disproportionate. The German trenches sited on reverse slopes simply could not be effectively dealt with despite the very best efforts of the artillery, and especially the mortars of 77e DI.[49] At the end of the day, after having failed again to make any serious progress, 97e RI was ordered to be relieved by elements of 109e Bde which had been sent to reinforce 77e DI. The order was given to consolidate any captured positions and begin connecting them with the rear. Some very small company-level attacks were launched on the 19th to stabilize the position of 77e DI and ensure liaison with the units on either side of it; but for all intents and purposes the battle was over.

A Tale of Two Divisions: Summary

There is no doubt that the experiences of 34e and 77e DIs were markedly different. They fought over different terrain, with different tactical objectives, and did so with sharply differing levels of logistical support. The composition of the two divisions was such that one was largely populated by elite, battle-hardened troops and highly skilled commanders, while the other was conspicuous for its sheer mediocrity in nearly every measurable index from élan to leadership. It is because of their differences rather than in spite of them that a study of these two units represents a more compelling case study than those that focus on a single 'representative' unit. Both divisions faced roughly similar problems. They both shared a common experience of the immense engineering work required to fight trench warfare, and the sagging morale which resulted from continuing on with

46 Ibid.

47 Ibid.

48 *SHD*, 24N1991; «*Compte-rendu des opérations de la 77e Division dans les Journées du 16 et 24 Juin 1915*» *25 juin 1915*.

49 Ibid. «*77e Division à 33e Corps d'Armée*» 18 *juin 1915*.

the mundane yet dangerous daily business of trench warfare after suffering heavy losses. To the men in the trenches these losses were not an abstraction, but were close friends and admired comrades lost forever. The return to routine trench maintenance and survival without the time to properly honour their lost comrades was incredibly difficult. In short, both divisions offer a glimpse into the soul and machinery of the French army in 1915.

When addressing the question of why 77ᵉ DI fared better than 34ᵉ, there is no simple, straightforward answer. Some would point to the overwhelming firepower at the disposal of 77ᵉ DI. Others would prefer to highlight the stunning leadership 77ᵉ DI enjoyed, even after the death of Barbot, while others still might choose to focus on more intangible qualities such as élan. It can be difficult at times to give any clear and irrefutable proof as to the efficacy or inefficiency of a given leader when left with nothing but archival evidence. The effects of things said (or not said), and in any case never recorded, are lost to the sands of time. Procedures in a large modern bureaucracy like the French army tend to be sufficiently formalized to make differences between commanders quite difficult to find, especially in this period when high-command dogma was so strongly thrust upon field commanders. What one is left looking for then are the smallest variations in wording, the regularity of contact (which itself can be deceptive as records frequently do not survive long enough to reach the archives), and any orders or actions which can be considered as being ‘above and beyond the call of duty’ or, at any rate, in more detail than is strictly required. The weighing of one leader against another is no less fallible than comparing one unit to another, and suffers many of the same pitfalls. However, a case can certainly be made that 77ᵉ DI did enjoy much better leadership than did 34ᵉ DI throughout Second Artois.

General Barbot had a keen mind for tactics and for the welfare of his men in the completion of their assigned mission. In many ways, he was an ideal divisional commander. His orders dealing with how troops were to advance on 9 May were given with great clarity and undoubtedly helped the assault battalions better perform the tasks assigned to them. His provision that assault troops only take forward with them materials that were directly relevant for conducting offensive operations – such as ammunition, barbed wire, food and water – lightened the load of the infantry. This not only allowed them to move more quickly, thus reducing their exposure to the German counter-barrage, but also staved off fatigue, which meant that the men would have more energy and be better able to resist the enemy throughout the duration of the engagement. Barbot’s detailed logistical organization of his front ensured a fluid deployment, supply and advance for his troops, who would not easily get lost in their own trenches or be otherwise impeded by troops coming back the other way down narrow *boyaux* and by other obstructions. While 77ᵉ DI would still suffer overcrowded *boyaux* filled with the dead and dying, Barbot’s strict regulation of the uses of the various *boyaux* undoubtedly reduced the confusion and delays which battle imposed on reserves marching forward into the fray.

His plan for the rolling barrage on 9 May was bold and innovative, and responded remarkably well to the specific conditions on his front. After *ouvrage* 123 the German defences thinned dramatically, which made *ouvrage* 123 a logical point at which the barrage could halt and wait for the advancing infantry. The artillery would be able to quickly and easily roll the barrage back if the infantry had been held up; and since the initial roll forward lasted only 10 minutes there was little chance of the barrage leaving the infantry hopelessly far behind and exposed to an organized German defence. Once the infantry had signalled to the artillery that they had reached *ouvrage* 123 the barrage continued to roll on to its subsequent objectives, thus providing protection for the men as they crossed the open land between *ouvrage* 123 and Souchez (even if the slow speed of the barrage meant that the men of 77ᵉ were nevertheless exposed to enemy fire during their advance). Barbot not only organized the barrage, but also instructed his infantry to follow as closely as they could (remembering that the barrage might have a 60–80m margin of error) and to try to hit German trenches before the Germans could organize any coherent defence. Such precise and practical advice was not often found in other divisions during Second Artois. Ultimately, it was a startling plan, and all the more so for having succeeded. The men of 77ᵉ DI met remarkably little German resistance until reaching Souchez, Givenchy and Vimy Ridge.

This contrasts sharply with the experience of 34ᵉ DI who, if they did fire a rolling barrage on 9 May, could not profit from it in any meaningful way. Whereas Barbot was a hands-on commander and played an inimitable part in the moulding of his division and the planning and execution of the attack, de Lobit seemed little more than a mouthpiece for his corps commander, Dumas. It was Dumas who organized the artillery barrage, gave instructions to the infantry concerning their responsibilities and generally organized the attack. To his credit, Dumas was fairly detailed in his prescriptions, but he was also very formulaic: his orders on the number of breaches to be opened in the enemy wire feel as if they were being transcribed out of a field manual instead of dynamically adjusted to the actual situation at the front. Likewise, his plan of having multiple rest periods and short, sharp bombardments in the final minutes before H-hour on 9 May seems like nothing more than a textbook method for confusing the Germans, which in the event would prove utterly ineffective. De Lobit's only real responsibility was ensuring infantry–artillery liaison and designating specific troops for *nettoyage* duty – jobs assigned to regimental commanders in 77ᵉ DI.

While 34ᵉ DI did not have nearly as sophisticated a plan as 77ᵉ DI, events might easily have proven that they did not need one. Unfortunately for the men of 34ᵉ DI this was not the case, and the men of 88ᵉ RI paid for that failure with their lives. Was it a result of poor leadership? It is very difficult to say. 34ᵉ DI did not have access to anything like the quantities of modern guns available to 77ᵉ DI and, perhaps worse, had only a fraction of the shells. Neither a rolling barrage nor a more meticulous organization of 34ᵉ DI's trenches and *boyaux* would have helped the infantry cross the largely intact German wire opposite them, or pass through the hail of enfilade machine-gun and rifle fire which had stopped the 10.00 attack on 9 May dead. Once the German artillery began to respond all hope was lost.

There is room for doubt that 34ᵉ DI's limited number of confirmed rolling barrages during Second Artois was due to the oversight of its commanders. Instead it is possible, even likely, that the material constraints placed on the division simply made a rolling barrage too costly, especially as it would not drastically affect the division's ability to perform its primary function, which was to protect the flanks of the main attack. Perhaps, then, the primary requirement for success was the presence of modern, quick-firing heavy artillery and an ample supply of shells? The operations from 9 to 11 May would certainly support the idea. XXXIII CA had access to far more guns and shells than any other corps, and did correspondingly much better. However, XXXIII CA (excluding 70ᵉ DI) also attacked a significantly easier front than any other corps in Tenth Army, and had elite corps on either side of it providing crucial assistance and drawing away enemy fire. XXXIII CA clearly had a number of factors working in their favour. Nevertheless, 77ᵉ DI was stymied once it reached solid German defences in and around Souchez in the same way that XXI and XX CAs were on Notre Dame de Lorette and in Neuville, respectively.

While the presence or absence of sufficient guns and ammunition was a key component of success, the period after 11 May typically saw 77ᵉ DI doing little better than 34ᵉ DI despite its material wealth. Facing tough German emplacements in Souchez the 77ᵉ could do little. To loosely paraphrase General Charles Mangin: 'no matter what they did it was going to result in heavy casualties'. This was especially true from 16 to 18 June when 77ᵉ DI, and notably 159ᵉ RI, failed to make anything but the most meagre progress. Indeed, the operations from 16 to 18 June are telling when one considers the details closely. The story of 34ᵉ DI is one of a division which, despite painstakingly re-learning the basic skills of trench combat after their utter defeat in the opening phase of the attack, got cut down by an efficient and effective German defence system which included flanking positions with wide interlocking fields of fire and pre-planned counter-barrages. Faced with overwhelming enemy fire, the vast majority of 34ᵉ DI's soldiers very quickly returned to their trenches. Some units made it to the German wire and, upon finding it intact, were compelled to retire, while a very small percentage of the division actually managed to enter, and temporarily hold, the German front-line trench.

While 77ᵉ DI faced many of the same difficulties, its men were far more likely to doggedly continue the advance, find a way into the German trenches and then advance as far as they could before running out of grenades, at which point a defensive position was formed in whatever captured German trenches could be maintained. While the favourable advance of 97ᵉ RI and its control of the Souchez cemetery from 16 June undoubtedly put the division in a stronger position than 34ᵉ DI, it did not seriously alter the situation facing many assault troops of the division when they were ordered to attack German trenches which sat comfortably behind thick and largely intact wire emplacements. What pushed the men of 77ᵉ DI to continue to advance under conditions which would send their comrades in

34e DI scrambling for safety was their élan. The division was made up of driven and aggressive soldiers and inculcated into its men an *esprit de corps* which predisposed them to press on despite having sustained heavy casualties: proof that there was some logic to the maligned pre-war theories of élan *vitale*.

This aggressive and hard-working spirit was not merely a result of good leadership, but also a reflection of the ideology of the men and lower-level units. This is most strikingly evident in the case of 67e Bde. 67e Bde, organically part of 34e DI, spent most of Second Artois fighting under 77e DI command. As such it provides an interesting, if not imperfect, case study in and of itself. During its time at the front 67e Bde was responsible for reorganizing the captured German trenches, uniting them with the rear and launching several small, local attacks. While the results of 67e Brigade's toil were not always laudable (Pétain thought that work was progressing too slowly after 11 May in all of XXXIII CA), the brigade did work at an accelerated pace for weeks on end. Even rest periods consisted of spending the night working on trenches alongside 88e Bde, which was holding the front. The discipline and élan of 77e DI seems however to have not remained with 67e Bde once it returned to 34e DI. While some small sections of 67e Bde front fared better than the 68e (their wire was much more effectively sited, for example), they had the same problems with indiscipline, especially concerning their lax approach to guard duty. During the attacks from 16 to 18 June units from 67e Bde seemed to have pushed harder and were less likely to abandon an advance at the first German retaliation. This, however, may have been due to nuanced variations of 67e Bde's front which do not come out in surviving records. In any case the internal spirit of 67e Bde, somewhat lazy and unprofessional, could not be permanently supplanted by the more effective ideologies of 77e DI in such a short period of time.

The differences in 34e and 77e DIs' methodology and ideology go a long way to explaining why the two divisions met with such differing results during Second Artois. However, what does the experience of these divisions say about infantry combat in 1915? First and foremost, it tells us that the division was the tactical unit in 1915. Excepting extreme cases like that of Dumas' XVII CA, corps commanders provided only a basic level of oversight when it came to infantry tactics. Corps command still had a fair amount of power owing to its allotment of heavy batteries; but in situations where there was sufficient firepower (such as in XXXIII CA's sector) a division could, in theory, make an attack without seeking assistance from any larger body. As such, the influence of the divisional commander was paramount, and could dictate the tone of an entire action. Under Pillot, 67e Bde was forced to work hard, but once the brigade returned to 34e DI, which was led by the relatively weak de Lobit, the unit seemed to lose a significant portion of its vigour. It was also within the power of the divisional commander to delve into the specifics of infantry combat and activity, ensuring a level of competence and understanding from his men. In weak units this competence was very low (one is reminded of Dumas walking the trenches to find soldiers utterly ignorant of the chain of command), which often made it unlikely for the men to perform great feats of either engineering or valour.

Also evident is the shift, already manifest in Second Artois, away from units being composed of nothing but identically equipped riflemen. As soon as the German defence stiffened after 9 May the single most important trench weapon *par excellence* became the grenade. Specially trained grenadiers led the way, and were responsible for a quantity of infantry advances sharply out of proportion to their numbers. *Nettoyeurs* had become an obvious part of any serious attack, and care was taken in their selection, supply and training. As German lines were increasingly sited in difficult to hit locations, as at Souchez and Chantecler, infantry-manned mortars became the primary arm for demolition fire – even more crucial than heavy artillery in some sectors. The infantry were called on more and more to take an active role in artillery spotting, which would eventually mean sound-ranging and flash-spotting as the war progressed.

As important as the increased specialization of infantry is to the development of modern warfare, the truly ascendant branch of the military machine in the First World War was the engineers. In 77ᵉ DI the engineers were used as planners and project managers, leading vast teams of infantry in the construction and development of their trench systems. They were also quartermasters, gas specialists, logistics organizers and administrators (it was the engineers who organized the supply of grenades for 34ᵉ DI, for example) and combat personnel. The First World War remains the pre-eminent trench war, arguably the largest siege in all of history, and in hindsight it should be no surprise that engineers would prove to be of paramount importance. Hand in hand with the larger logistical apparatus, engineers were the oil that kept the war machine running. The excessive difficulties faced by 77ᵉ DI after 11 May in consolidating its position and digging acceptable *boyaux*, and the strains that the lack of an effective trench system placed upon the division seriously eroded 77ᵉ DI's battle efficacy. Despite a huge effort ordered by Pillot, and directed by his engineers, the division still suffered remarkably poor communications and living conditions, even in the rear, after five weeks of constant digging. As important and interesting as are the permutations of infantry in the First World War, they are meaningless without a well-organized trench system designed by engineers; a well-sited artillery bombardment to neutralize the enemy defences; and strong leadership to mould, train and direct the division toward its goal – the ultimate goal of the Allies on the Western Front: to defeat the German army in the field.

Conclusion
The Impact of Second Artois

Hans Delbrück, the father of modern military history, wrote that warfare is the single most complex human endeavour; so complex, in fact, that man can never comprehend all of the myriad variables that play a part. This was true in antiquity and it is true now. Warfare experienced a sharp increase in the number of variables involved during the First World War: all the more reason for historians to be drawn to 'seminal moments', like the Somme, in their attempt to make sense of it all. Still, the benefit of hindsight allows historians to piece together far more data than even the most well-informed staff officer or field commander, and now we might begin to piece together some of the rich and complex periods of the war which have hitherto been overlooked.

Despite its importance, Second Artois seems to have been all but forgotten, despite its being well memorialized. Statues and monuments abound in the region, including an increasingly derelict monument atop Vimy Ridge dedicated to the Moroccan Division, and the vast ossuary and chapel of Notre Dame de Lorette. Nevertheless, modern historians have only a few small excerpts from various history books with which to piece together the entire battle; without the French Official History or Palat's history of the war (written in the 1920s) even a modest understanding of the battle is unattainable. Those without a working knowledge of the French language may never come across a suitable narrative of the battle. The lack of research on Second Artois, and most other battles from 1915, has obscured their vital importance to the development of the war. Unfortunately, by merely engaging with what little information is available to historians outside the archives, Second Artois is too often quickly categorized as 'just another failure' and then tossed aside along with a valuable opportunity to learn a great deal about the First World War.[1]

Could Tenth Army have succeeded in capturing Vimy Ridge in spring 1915? Undoubtedly – although the likelihood of a 'continuous' battle yielding such a result is dubious. As long as the Germans held Souchez and Neuville, Vimy Ridge was safe; and the complexity of the German positions in either village was such that there was almost no chance of them being captured in a single bound. The only sound plan available was Foch's original vision for the battle laid down in March 1915. Tailoring the operation to the available materiel on hand, as Foch's *longue durée*, three-step approach was intended to do, was the only way of

1 Clayton, *Paths of* Glory, pp. 70–71 and Doughty, *Pyrrhic Victory*, p. 165.

ensuring that each attack would be launched with a sufficient concentration of men and materiel: a brief look at the experience of units such as 34ᵉ DI shows the hopelessness of attacking without adequate artillery. Even so, the operations would not have been easy, and would still have resulted in a heavy loss of life. One must also expect that, faced with French success in capturing either Souchez or Neuville, the Germans would have worked hard to protect what they had left and develop defences which would thwart future French attacks, just as they did after the French advance on 9 May.

Despite its size and importance the Second Battle of Artois remained at the forefront of French thinking for only three months. During those three months, however, the actions of Tenth Army from 9 May to 18 June would reverberate strongly across the entirety of the French army, and greatly shape the planning for the French offensives on 25 September 1915 (Third Artois and Second Champagne). Despite the ultimate failure of Tenth Army to secure Vimy Ridge many in the French high command felt that Second Artois (especially the actions on 9 May) was proof of the French army's ability to win great tactical and potentially even strategic victories on the Western Front over the course of a day or two's fighting. This confidence was combined with a steady increase in materiel which would allow the French to launch multiple full-scale attacks simultaneously for the first time since the beginning of trench warfare on the Western Front.[2] From the end of Second Artois to the launching of Third Artois and Second Champagne Joffre led a vigorous effort to secure the necessary materials needed to launch successful trench offensives. In June he got the War Ministry to commit to doubling the number of machine guns per brigade (an increase of some 5,500 machine guns overall) by 1 January 1916. Joffre also oversaw the increased production of trench guns, including the modern 240T and 340T mortars, as well as the production of the first *artillerie lourde à grande puissance* (ALGP) batteries, which could strike deep into the German rear – an important response to the German use of super-heavy guns to fire on important allied rear areas like Dunkirk.[3] The production of 75mm shells was actually slowed down to reduce the problematic quantity of malfunctioning shells which had been produced since the beginning of the war, and which had so greatly taxed the guns used during First Champagne and Second Artois. Gas shells and aircraft were both highly praised for their effective contributions to Second Artois, and large orders were placed for the rapid expansion of both arms.[4]

With modern weapons reaching the front lines in ever-increasing numbers, high morale and increasing diplomatic pressure from the Russians, who hoped that the French would draw German pressure away from the Tsar's beleaguered forces, the French felt compelled to launch another large-scale attack as soon as possible after the end of Second Artois. In late June 1915 Joffre consulted the commanders of the

---

2 *AFGG, Tome III*, pp. 151–152.

3 Ibid., pp. 154–157.

4 Ibid., pp. 157–158.

newly created North and Centre army groups[5] (Foch and Castelnau, respectively) as well as Pétain, the newly appointed commander of Second Army, and asked them to propose a location, objectives and methodology for the French army's next attack.[6] The three men did not delay long in giving Joffre their responses, each of which expressed a different interpretation of Second Artois. Foch asserted that the French did not yet have sufficient materiel (especially in terms of poison gas and heavy artillery) to risk a large-scale attack whose sole function would be to break the enemy front. Instead of this vague and imprecise objective Foch asserted that the French ought to 'fix for each one of our attacks a goal, an objective likely to be attained'.[7] He felt that 'in sum, an attack launched without a precise and detailed objective was incapable of producing results, especially a breakthrough.'[8] Foch expressed his opinion that Castelnau's sector (Champagne) offered nothing in the way of important military objectives, whereas a further attack in Artois could yield crucial pieces of terrain such as Vimy Ridge, the possession of which would greatly improve the situation of French forces along the northern flank.

Castelnau felt differently. He explained to Joffre his belief that 'any offensive action against a relatively large enemy front ... should have as its initial goal the conquest, in a single and vigorous bound, of the entire depth of an enemy defensive position.'[9] As such, Castelnau felt that areas in which German defences were weaker offered the greatest chance of success: areas such as Champagne, which was in the sector of Castelnau's *Groupe des Armées du Centre* (GAC). While it is perhaps tempting to assume that both commanders were doing whatever they could to ensure that new supplies of men and materiel went to their army group as opposed to someone else's, it is important to consider how the two men conceived of operations. Foch, since the onset of trench warfare, had championed a slower, more methodical approach to battle, while Castelnau believed in the 'continuous battle' whereby a steady stream of reinforcements would provide sufficient impetus to carry French forces through the entirety of the German trench network. The extent to which the different sectors the two men found themselves in (the industrial, populated north or the gently undulating and underpopulated Champagne region), or their branch (Foch was an artilleryman, Castelnau an infantryman) influenced their opinions is open to debate. Ultimately,

[5] On 18 June 1915 Joffre split the Western Front into three army groups: North under Foch (being simply the old *Groupe Provisoire du Nord*, GPN); Centre under de Castelnau; and East commanded by General Dubail.

[6] *AFGG, Tome III*, p. 178.

[7] Ibid., p. 179: «fixer à *chacune de nos attaques un but, un objectif susceptible de rendement*».

[8] Ibid. «En résumé, *une attaque lancée dans l'espace sans objectif pris et défini est incapable de résultat, de* percée *en particulier.*»

[9] *AFGG, Tome III*, p. 181: «*toute action offensive contre une portion plus ou moins* étendue *de front ennemi ... . Son but initial doit* être *de conquérir d'un seul et vigoureux élan et dans toute sa profondeur le système défensif organisé par l'ennemi.*»

what is important is that both men took very different lessons away from Second Artois. Foch saw the failure of Tenth Army to capture Vimy Ridge as a result of d'Urbal's having opted for a large, brusque attack instead of undertaking a lengthy 'operation' in which key enemy strong-points were taken individually until Vimy Ridge could be safely assaulted. Castelnau saw the advance of 9 May 1915 as evidence that given sufficient men, materiel and the benefit of surprise, the French could easily smash through any German trench network.

Foch's position was largely seconded by Pétain, who also felt that a methodical approach was more appropriate, and that Vimy Ridge was probably the single most important piece of terrain the French army could hope to capture at that time.[10] Pétain was less convinced, however, that Tenth Army would be in a position to launch another major assault on Vimy Ridge any time soon. D'Urbal was still having difficulties subduing Souchez, and the men of Tenth Army were tired and increasingly demoralized. Furthermore, the proposed plan to place Pétain's Second Army to the right of Tenth Army in the hopes of having it launch a major supportive attack did not seem to promise sufficient tangible benefits to the operations north of Arras. Pétain felt that while attacking in a high-value area such as that north of Arras would be a slow, difficult and costly affair, there was no place along the Western Front where the Germans could not rely on well-fortified villages, woods and trench networks. In response to this situation, Pétain suggested what the British would later call 'bite and hold' tactics: a series of small attacks with limited objectives, which would capitalize on the French maintenance of the initiative on the Western Front to attack with surprise and local superiority in men and munitions. He explained that:

> These partial attacks do not all have to have the same importance. Some of them may even transform into veritable battles, based on how the attack progresses. It is, in sum, decentralized battle: a series of small conflicts in which each commander would choose his terrain and his time, and exploit the vigour and confidence of his troops.[11]

Soon after having received the three men's suggestions, Joffre decided on 8 July that the next major attack would be launched in Champagne by de Langle de Cary's Fourth Army (which had fought in the first Champagne battle) and Pétain's Second Army (reinforced with elite units, such as the Moroccan Division and XX CA). D'Urbal's Tenth Army would launch a supporting attack in Artois a few days

[10] Ibid, p. 180.

[11] Ibid., p. 181: «*Ces combats partiels ne sont pas appelés à avoir tous la même importance. Quelques-uns même pourront se transformer en véritables batailles, grâce à fléchit définitivement dans une zone. C'est en somme la bataille décentralisée en une série de combats partiels où chaque chef choisit son terrain et son heure, et exploite au mieux la vigueur et la confiance des troupes.*»

prior to draw in German reserves.[12] This strategic decision did not, however, stop the debate between Foch and Castelnau. At a meeting of the three army group commanders (Foch, Castelnau and Dubail) and the commander-in-chief, Foch and Castelnau again aired their divergent views of how operations should proceed on the Western Front.[13] Foch was still insistent that the war needed to be fought 'like a siege', and that only massed munitions and a methodical approach could bring success. Castelnau reiterated his belief that 'our goal is to create a flank in the enemy disposition and then manoeuvre against it'.[14]

Castelnau boasted that such a brusque attack would yield 10–12km in the first 24 to 48 hours, results which a methodical approach could never achieve. Joffre sided with Castelnau, stating his belief that Foch's proposed methodical battle 'would entail a month of combat, with a maximum expenditure of ammunition; at what point would we be able to declare ourselves ready to attack? Maybe not next year, and probably never. We must act, for us and for our allies. As our regulations say: only inaction is disgraceful.'[15] Foch's plan would consume quantities of munitions which the French could not spare, Joffre claimed, and was simply not fast or bold enough to take advantage of the Germans having moved yet more divisions to the East. Joffre felt that 'the hour has come for each to give it his all', and this strategic decision dominated tactical developments leading up to 25 September.[16]

Second Artois had seen tactical and logistical refinements thus far unparalleled on the Western Front. News of the achievements and methodology of Tenth Army spread quickly, owing to notes produced by *Grand Quartier Général* (GQG) as well as to liaison between various armies and army groups. De Langle de Cary had a staff officer from Fourth Army posted to Tenth Army during Second Artois who, by early July, had compiled a report on the lessons learned from Second Artois. While the note is incomplete (an unknown number of pages are missing), it provides an interesting glimpse into how some French soldiers outside the high command viewed the battle. The officer, whose name is unknown, confirmed the consensus that Second Artois was launched in accordance with Note 5779, and that the successes won by Tenth Army served to validate the prescriptions of that document.[17] According to the report (and many other similar documents) this

[12] The Artois attack was later moved forward to coincide with the Champagne attack as a result of various strategic, tactical and logistical considerations.

[13] *AFGG*, *Tome III*, p. 275.

[14] Ibid. «*notre but est de créer une aile dans la dispositif ennemi, puis de la manœuvrer*».

[15] Ibid. «qui *exigerait un mois de combat, avec des dépenses maxima de munitions, à quelle époque pourrions-nous nous déclarer prêts à attaquer? Peut-être pas l'année prochaine, sans doute jamais. Or il faut agir, pour nous et pour nos alliés. Comme disent nos règlements: l'inaction seule est infamante.*»

[16] Ibid., p. 277: «*l'heure est venue pour chacun de donner son rendement maximum*».

[17] *SHD*, 19N735; «*Renseignements recueillis sur place par un Officier de l'Etat-major de la IVe Armée sur les actions offensives de la Xe Armée dans le Nord» IV Armée, 6 juillet 1915.*

sentiment was felt more or less universally by Tenth Army soldiers, officers and observers alike.

The report devotes considerable time and effort to describing, in great detail, the ideal layout of a trench network to facilitate the effective and uninhibited flow of men and materiel. Specific tactical lessons from Second Artois are also mentioned, especially in terms of the importance of firepower. The necessity of having a very large quantity of grenades available is highlighted by the report's mentioning that 53ᵉ DI used 23,000 grenades in only two days.[18] Artillery firepower is also mentioned, but not simply in terms of numbers of shells and guns. The report emphasized the fact that artillery preparation needed to be not only long and intense but also very precise, and praised Pétain's meticulous attention to detail and constant liaison with his artillery, which together helped ensure the 'perfect' artillery preparations which the Moroccan Division benefitted from in both May and June.[19] When dealing with the nature of infantry tactics the report does not differ greatly from Note 5779 in the details; it does however seem to have a slight 'continuous battle' tinge.

The most prevalent theme of the infantry component of the report deals with maintaining the impetus of an attack for the purpose of conquering as large an area as possible in a single push. The XX and XXXIII CAs are consistently praised for their having taken the first two enemy positions on 9 May without stopping (the sub-par performance of XX CA's 11ᵉ DI is not mentioned).[20] Their ordering of assault waves to maintain their advance by bypassing enemy strong-points, combined with their use of *nettoyeurs*, is held up as exemplary. Also emphasized is their having infantry advance above the captured enemy *boyaux* to prevent the attack from being stalled by overcrowded or obstructed trenches, while also maintaining the offensive spirit or 'élan' of the men (although the word is not used in that context in the report). The report details the organization of the infantry waves, and the extensive amount of preparatory digging that facilitated the quick forward movement of reserves in both corps; each division in XX CA was reported as having dug 25–27km of *boyaux* during Second Artois.[21]

The detailed trench systems which helped to facilitate the rapid forward deployment of assault and reserve troops in XX and XXXIII CAs, combined with their effective training and use of assault and *nettoyeur* troops, are held up as the only way to ensure 'results'. This representation of 9 May seems to fit in very well with the popular sentiment after Second Artois (repeated in the report) that had the reserves only been able to get forward faster XXXIII CA might have

18 Ibid.

19 Ibid.

20 Ibid.

21 *SHD*, 19N735; «*Renseignements recueillis sur place par un Officier de l'Etat-major de la IVᵉ Armée sur les actions offensives de la Xe Armée dans le Nord*» IV *Armée, 6 juillet 1915.*

retained Vimy Ridge.[22] While a few practical problems with making such a large advance are discussed in the reports written after Second Artois (such as the great difficulty in ensuring that forward troops have sufficient fresh water to drink), they are mostly addressed at a minor tactical level without consideration of wider tactical or operational issues which arise when some units advance deep into enemy positions while units on either side of them fail to do so.

Pétain himself, in his new capacity as commander of Second Army, did not remain idle but instead produced tactical reflections of his own between 18 June and 25 September. The common thread throughout them is a reaffirmation of the efficacy of the provisions laid out by Note 5779. Instead of replacing the doctrine laid out in April, Pétain's notes tended to refine and add detail to the doctrine, more than anything. Much of Pétain's writing dealt with the ideal construction of a trench network: the number and location of trenches, *boyaux* and *parallèles de départ* as well as the number of men and man-hours required to build such networks.[23] The main goals of his proposed method of organizing trench networks were to allow all attacking troops to advance simultaneously, and to provide protection for reserve troops as they marched into battle. Pétain's influence was not limited to Second Army but spread through the entirety of the GAC, owing to his position as the adjunct to Castelnau: a position which afforded him considerable influence in the command structure of the GAC (there was a special *groupement* of artillery directly under Pétain's command, for example). Pétain was influential in writing the op-order for the entire GAC for 25 September 1915. The plan was as detailed and clear as an army group-level order could be, and provided an excellent designation of sectors, goals and resources for each of the six corps involved in the attack.[24] It did, however, conform to the strategic goals dictated by Joffre and Castelnau, which meant that cavalry was to stand by, ready to charge into the enemy rear areas where they would disrupt German transportation (railroads, primarily) and command structures. Second Champagne was to be a breakthrough battle.

While Second Champagne would be conducted as a breakthrough battle, it would not be launched in the hope of a return to mobile warfare – as the British army might have. Rather, the hope was to force the Germans to retire out of the large salient they held, which pointed towards Paris, by launching attacks on the two wings of the salient (at Artois and Champagne). While the details of these battles are far beyond the remit of this book, the failure of the attacks on 25 September to meet their goals are important. Despite the success of Second Army (which captured some 4 kilometres), and despite the significant strain which these battles (in addition to their analogues, such as Loos) placed on the German army, it became clear to many

22 Ibid.

23 Ibid. «Note relative à la *préparation d'un terrain d'attaque en vue de l'action de l'infanterie*» *IIe Armée, 28 juillet 1915.*

24 *SHD*, 19N735; «Instruction *préparatoire aux opérations de la Nème armée*» A.G.A.C., 6 *août 1915.*

in the French army that 'brusque' attacks, breakthrough battles, were not going to yield the vast strategic gains that they hoped for.

While the battles of 25 September 1915 were important, they did not carry with them the same long-lasting doctrinal ramifications as Second Artois, whose conduct was still the basis of important works of doctrine into 1916, if not later (Foch's *bataille scientifique* and his large doctrinal note *La Bataille Offensive* from 1916 are notable for their strong correlation to Note 5779 and the refined lessons of Second Artois). The monumental efforts of 25 September 1915 were very quickly overshadowed by events at Verdun and along the Somme: two battles so crucially important to the French and British nations that they still serve as the iconic examples of combat in the Great War, taking precedence even over the Allies' march to victory in 1918, of which few members of the public can name a single engagement. This being the case, it is not at all expected that this book can reverse nearly 100 years of social memory and mourning.

Was 1915 a year of unmitigated disaster, as Clayton and Porch would have us believe?[25] Even if we conclude that it was a disaster, is this the result of the uncaring, callous and dangerously ignorant commanders of popular memory, wantonly ordering their men over the top? Few things are ever set in stone, but at least a handful of uncontroversial conclusions might be offered. 1915 cannot, in any reasonable way, be painted as a successful year for the Allies, which is perhaps why it receives so little attention from historians seeking to highlight the more successful facets of the Allied war effort. From the Dardanelles to Picardy, French and British attacks consistently failed to gain their objectives and suffered grievous losses in the attempt, which typically (but not always) far outstripped the losses inflicted on their enemies. Even that most ephemeral objective of 'supporting Russia' cannot really be said to have been accomplished; at any rate it would be very hard to prove that any of the Western Front battles in 1915 had a positive impact on the Eastern Front.[26] It is not surprising, based on the outcome of the majority of 1915 battles, that Falkenhayn should be convinced that he could win a genuinely attritional victory at Verdun; after all, the Germans had done so for the better part of a year already.

If Clayton and Porch are correct, and 1915 was a year the Allies would rather forget, their assertion that the repeated failures of that year were the result of poor French leadership could not be further from the mark. Anything more than a cursory glance at operations in 1915 shows just how engaged and intelligent most of the French command was. This is not to say that every French general was an effective field commander or administrator; nor does it imply that because many French officers were well meaning and intelligent professionals that they were free from making mistakes. It merely suggests that the French command system self-consciously endeavoured to improve at every level, and that French officers, almost universally, cared very deeply about the lives and well-being of their

25 Clayton, *Paths of Glory*, p. 60; Millett and Murray, *Military Effectiveness*, pp. 208–212.

26 Doughty, *Pyrrhic Victory*, pp. 160–161 and 168.

men. Note 5779, in and of itself, should speak volumes on behalf of the French command system, and seriously challenges Bruce Gudmundsson's assertion that 'centralization prevented the British and French Armies from adapting quickly to new situations'.[27] The French high command, and even GQG's 'Young Turks', took part in a process by which lessons learned at the front were collected, collated, processed and then distributed throughout the army. Over-simplistic explanations which demand that the learning process fit neatly into one of only two categories (top-down or bottom-up) do little justice to the much more organic, living system by which the French army evolved and improved as the war progressed.[28] Instead of acting the role of the dogmatic preacher forcing doctrine down the throats of infantrymen who would promptly spend six days of the week ignoring Sunday's sermon, GQG acted as a pump, intelligently circulating good ideas and encouraging the replacement of less than good ideas.

During the process of doctrine creation (really aggregation) and dissemination GQG could not be purist, but was required to fulfil its roles as the director of strategy and manager of human resources, as well as the distributor of tactical precepts. This, more than anything, seems to explain the existence of over-idealistic breakthrough rhetoric in Note 5779. It was GQG's job to encourage the men, and to hold before them tangible, significant rewards for victory, like the breaking of the trench network and a return to mobile warfare. When leading a large body of human individuals the general or politician in charge will always say certain things that no one (or almost no one), including the speaker himself, will believe; nevertheless they absolutely must be said, for the psychological effect of not saying them is simply too damaging. If a soldier cannot look forward to any noticeable progress he will quickly despair and conclude, as modern interpretation seems to have done, that the many attacks he is being asked to carry out, and perhaps the entire war itself, are ultimately pointless.

Undoubtedly, there were officers at GQG and, indeed, at all levels of command who wholeheartedly believed in the *percée*. However, their view did not represent the majority view of the nature of operations on the Western Front, as has sometimes been claimed. Joffre, engineer that he was, saw the war as a giant siege.[29] Haig, being a cavalryman, saw the war as one massive battle spanning years which, as with the great battles of the nineteenth century, would end in an exploitation phase where the losing side would shed territory and casualties at a very rapid rate (undoubtedly with cavalry chasing at their heels). While a tactical breakthrough of sorts was implied in Haig's vision it did not necessitate any end to trench warfare and, interestingly enough, was largely how the last three months of the war played out, with the Germans losing huge tracts of land and hundreds of thousands of casualties in the face of the Allied onslaught.

27 Gudmundsson, *Stormtroop Tactics*, p. 173.

28 Ibid. and Griffith, *Battle Tactics on the Western Front*, pp. 26–27.

29 *SHD*, 16N1905; «*Entrevue avec Lord Kitchener*».

If one looks closely at the battles fought by the French army in 1915 not one of them has as its express purpose the breakthrough of the enemy trench network and the resumption of mobile warfare. When Foch first envisioned what would become Second Artois it was as a slow, multi-phase engagement whose sole purpose was to capture important pieces of high ground from which the French would continue to pursue offensive operations. The observation offered from Vimy Ridge over the Douai plain is stunning to this day, and would have afforded the French a huge advantage in that sector (a fact often noted by Foch). Even when d'Urbal changed the tactical nature of the battle from a *longue durée* operation to a brusque 'continuous' battle its strategic aims remained the same: the capture of Vimy Ridge to facilitate further operations along the northern flank of the Western Front. Likewise, the battles of 25 September 1915 had as their strategic objective the pinching out of a major German salient which pointed towards Paris by attacking its sides in Artois and Champagne. The hope was not to break the trench deadlock but to force the Germans, through manoeuvre, to make a large strategic retreat, thus winning back a significant portion of occupied France. British battles on the Western Front (Auber Ridge, Festubert and Loos) were fought to support French efforts elsewhere.

Tactically, French doctrine in 1915 was also well suited to the situation at the front, even if it would still be greatly refined as the war progressed. The rolling barrage would be an essential tool throughout the war, and the early adoption of this crucial tactic by not only French artillerymen but by GQG must certainly be counted as one of their more notable early insights. Its efficacy is best seen in the advance of 77^e^ DI on 9 May when the rolling barrage enabled the advancing infantry to tackle a series of German positions, one after the other, without the Germans having sufficient time to man their machine guns and organize a defence. The rather vague terms in which Note 5779 prescribed the barrage might be seen as a good thing or a bad thing, depending on one's point of view; but, if nothing else, it did allow a large amount of room for improvisation and the fine-tuning of rolling barrages to local conditions: for example, the stopping barrage of 77^e^ DI that patiently waited for the infantry to reach *ouvrage* 123 on 9 May, its planned fanning-out barrage in June, and the hybrid barrage Tenth Army fired on 16 June. Such innovation could not have resulted from stodgy conservative minds reluctant to change. Likewise, the ready adoption of gas shells, aerial reconnaissance and on-board radio (TSF) shows an eager willingness to evolve and meet the problems which faced the French artillery.

As it is, the use of gas shells, mobile TSF receptors and the rolling barrage in spring 1915 is largely unknown, and almost always elicits surprise even among scholars. The use of artillery in 1915 is seen as being too often naïve and unscientific. French bombardments lasting for days or weeks are condemned for sacrificing even the slightest hope of achieving surprise.[30] Such condemnations seem to take no account of the strict material limitations within which the French

[30] Millett and Murray, *Military Effectiveness*, p. 212; Palat, *La grande guerre sur le front occidental*, p. 237.

army had to operate in 1915. They also seem to ignore the priorities of an artillery bombardment, of which destroying the enemy wire was always first. As the attacks of 16 June show – especially the attacks of IX and XX CAs after launching gas shells into La Folie) – no matter how sophisticated and effective the many component parts of an artillery preparation might be (counter-battery fire, barrages of varied description, neutralizing fire), if the enemy wire remained sufficiently intact then any infantry assault will necessarily falter and fail. No amount of surprise was going to catch *chevaux de frise* and steel wire asleep and unawares; the only way through an enemy wire emplacement was paved by heavy explosives, of which artillery guns and mortars were the best delivery system available in the First World War. Based on the experiences of 1915 (especially in September) one should expect the French to be increasingly cautious, and willing to engage in a days' or even weeks' long struggle for artillery dominance before attempting an infantry attack. While this tactical methodology was far from ideal, it was responding to very real material concerns, and as such needs to be understood as operating within a larger logistical, operational and strategic framework (complexities not always addressed by armchair generals and popular histories).

For the infantry, 1915 represented an equally transitional year. As the war became increasingly complex the infantry found themselves facing greater dangers, but also enjoying greater refinement. As bombardments and other offensive tactics became increasingly effective defensive tactics responded by become deeper and stronger; German bunkers lit by electric lights begin to crop up along the front in mid-1915. For the French, who were not as likely to develop overly complex defensive positions owing to their belief that upcoming attacks would carry them forward into enemy trenches, thus abandoning any feats of engineering they left behind, even relatively inactive periods became increasingly dangerous. Once the Germans had been awakened in the Artois sector they gave the French little respite from their sustained artillery bombardment, which chipped away at French morale and manpower. As trench systems grew they also became increasingly unsanitary, with human waste posing a serious threat to the fighting capacity of the infantry. With the deepening of trench networks and the intensifying of daily German bombardments soldiers could not relieve themselves in normal latrines, and thus had to develop means of isolating, disinfecting and disposing of large quantities of human waste: not an easy or pleasant task, as the experiences of 34[e] and 77[e] DIs attest.

On the attack French infantry found itself growing ever more powerful as the war progressed. By mid-1915 a French infantry division going on the offensive would likely carry forward with it light machine guns (British Lewis guns); various grenades, including British Mills bombs and Bessozi gas grenades; Bangalore torpedoes, petards and other assorted explosives, all carried by engineers; heavy machine guns, barbed wire and sandbags to repair captured trenches for defence; a small number of mortars, flags, flares, telephone equipment and, increasingly, gas masks; all in addition to their weapons, ammunition and extra rations. Despite the majority of popular attention being given to the development of tanks, aircraft and (on occasion) artillery the infantry went through a dramatic revolution of its own, through which men came to resemble something more like today's infantry specialists rather than their Napoleonic forebears. This is to say nothing of the burgeoning bureaucracy which supported the fighting men, and the sophistication of logistics and management reached by war's end.

How significant were the military developments in 1915? Was this a year of crass errors and blind alleys, or one of great intellectual fertility? It would seem logical that when faced with a new problem (i.e. the trenches) the armies on the Western Front would try to adapt as quickly as possible and, as such, one is likely to see the most rapid period of change shortly after the problem is presented. In many ways 1915 bears this out. While there were significant innovations and improvements to methodology that took place in 1916–18, the vast majority can trace their lineage back to 1915. The rolling barrage, infiltration tactics, sound-ranging, defence-in-depth and a long list of other key First World War developments were not part of the natural evolution of European armies as they finally severed themselves from their Enlightenment past and embraced the Industrial present. These tactics and practices were developed for the express purpose of tackling trench warfare. While many would go on to be further developed and modernized (Tukhachevsy's 'deep battle' and Nazi Germany's 'blitzkrieg' probably being the most visible examples of how First World War ideas continued to evolve after the war), their birth is nevertheless tied inextricably to trench warfare and, thus, to the early trench battles of 1915.

It would be unreasonable to expect the French army to have quickly mastered the many new offensive arts it developed in 1915. The French were capable of great tactical feats (9 May 1915), but were not yet sufficiently experienced or provisioned to perform at that level consistently. To make matters significantly more difficult the Germans were innovating at the same time, which meant that no solution that the French could invent would permanently 'solve the problem'. The best method was not to find the winning recipe first but, instead, to continuously out-innovate the enemy, and to press home an advantage as hard as possible before the opposing force could find a way to counter it. Such a vibrant, living, intellectual and physical struggle took place against a backdrop of logistical and industrial competition, the strict management of manpower and global finance. The first faltering steps of the French army in the application of its new offensive doctrine in 1915 are as important as the successes on the Somme and at Verdun, if not more so.

How does an understanding of 1915 on the Western Front contribute to our wider understanding of the First World War? The sheer lack of attention given to it makes it fascinating. This is a result not only of the fact that 1915 is still often viewed as a year of failures (and thus not fitting in well with the 'learning curve' concept, or at least not being the interesting part), but also a result of the fact that the nation perhaps most infatuated with the war, Great Britain, performed a relatively limited role in it. This work has already shown glimpses of the promise which research into the early trench battles holds. 1915 may yet conceal a few elusive hominins linking the anatomically modern armies of 1918 with their simian past. If one reads Foch's 1916 doctrinal work *La Bataille Offensive*, one finds a doctrine which is clearly a part of the same intellectual evolution as Note 5779. The battles of 25 September 1915 cannot be fully understood without looking back at Second Artois and appreciating the French insistence on its ability to force a major strategic result by a single, abrupt tactical victory. Tactically, the relations are clearer yet. Units in both Artois and Champagne planned their attack (and especially the placement of their reserves) based heavily on the experiences of Tenth Army in Second Artois.[31] Indeed, Joffre's entire plan for 25 September 1915 exhibits a continued belief in the possibility of a *percée* forcing a major German withdrawal.[32] The ultimate failure of 25 September would then inform how the French conducted the big 1916 battles (Verdun and the Somme), both of which retained large elements of the tactical and operational lessons learned in early 1915.[33] Thus, to study 1915 is to study an important chapter in the four-year-long evolution of trench warfare on the Western Front, and to understand the fullness of that evolution a little bit better.

---

[31] *AFGG, Tome II, Annex II*, p. 1076 and Guiral, Louis (1965), *'Je les grignote ...': Champagne 1914–1915* (Paris: Hachette), p. 123.

[32] Doughty, *Pyrrhic Victory*, p. 189.

[33] Philpott, *Bloody Victory*, p. 96.

Appendix

Tenth Army Order of Battle

III CA (Corps d'Armée
5e, 6e DI *Division d'Infanterie*)
239e, 274e RI (*régiment d'infanterie*)
IX CA
17e, 18e DI
268e, 290e RI
141e RIT (*régiment d'infanterie territoriale)*

X CA
19e, 20e DI
241e, 270e RI

XVII CA
33e, 34e DI
207e, 209e RI

XX CA
11e, 39e DI
142e RIT

XXI CA
13e DI, 43e, 48e DI
57e, 60e, 61e BCP (*bataillons de chasseurs à pied*)
$143^{e,}$ 144e RIT

XXXIII CA
70e, 77e, DI
Division Marocaine

Additional Divisions
51e, 53e, 55e, 153e DI

Bibliography

Primary Sources

Laffargue, André (1916), Étude *sur l'attaque dans la période actuelle de la guerre: impressions et réflexions d'un commandant de compagnie* (Paris: Plon-Nourrit et Cie).

Décret du 28 octobre 1913 portant règlement sur le conduite des grandes unités (service des armées en campagne) (Paris, 1913).

Règlement provisoire de manoeuvre de l'artillerie de campagne (Paris, 1902).

National Archives: War Office (WO)

WO 158/26 – Anglo-French relations in spring 1915

Liddell Hart Centre for Military Archives (LHCMA), King's College London

Robertson 4/1 – papers concerning 1915

Hamilton 5/1/7 – papers concerning the planning of Gallipoli and reports on the French conduct of battle in early 1915

Service Historique de la Défense (SHD), Château de Vincennes

16N707 – Gas shells
16N826 – Poison gas
16N832 – Poison gas (mostly protection)
16N1807 – GQG letters received from the GPN
16N1677 – GQG letters and doctrine
16N1905 – GQG correspondence with the British army
16N2095 – Artillery notes, reports and doctrine 1915–18
17N303 – Military mission
18N94 – GPN correspondence with GQG
18N134 – GPN correspondence with the British army
18N148 – GPN doctrine (especially, *La Bataille Offensive*)
19N735 – Offensive works before Second Champagne
19N1677 – GQG instructions and doctrine
19N1685 – Tenth Army artillery
19N1686 – Tenth Army heavy artillery
22N163 – IV CA documents from First Champagne

22N573 – IX CA from 29 May to 25 June 1915
22N1357 – XX CA, May 1915
22N1358 – XX CA, June 1915
22N1457 – XXI CA from 10 to 26 June 1915
22N1471 – XXI CA during First Artois (Notre Dame de Lorette)
22N1472 – XXI CA during Second Artois
22N1500 – XXI CA artillery in 1915
22N1808 – XXXIII CA correspondence from 11 May to 18 June 1915
22N1832 – XXXIII CA in 1915
24N214 – 11e DI during Second Artois
24N741 – 34e DI during Second Artois
24N1754 – 70e DI during Second Artois
24N1991 – 77e DI during Second Artois
24N2908 – Moroccan Division during Second Artois
25N87 – 104e RI during First Champagne
25N172 – 159e RI during Second Artois
26N163/3 – XVII CA *Journal des Marches et Opérations* (JMO)
26N326/2 – 34e DI JMO
26N407/2 – 77e DI JMO
26N667/12 – 88e RI JMO
26N701/1 – 159e RI JMO
Microfilm 9M394 – 104e RI JMO

Secondary Sources

General Accounts of Operations and Doctrinal Evolution

Campana, J. (1923), *Les progrès de l'artillerie: l'artillerie française pendant la guerre 1914–1918* (Paris: Imprimerie la Renaissance).

Clark, Alan (1962), *The Donkeys* (London: Hutchinson).

Clayton, Anthony (2003), *Paths of Glory: The French Army, 1914–1918* (London: Cassell Military).

Daille, Général M. (1936), *Histoire de la Guerre Mondiale: Joffre et la guerre d'usure 1915–1916* (Paris: Payot).

Doughty, Robert (2005), *Pyrrhic Victory: French Strategy and Operations in the Great War* (Cambridge, MA: Belknap Press of Harvard University Press).

Dubois, Pierre Joseph Louis Alfred (1916), *L'Artillerie de campagne dans la Guerre Actuelle 75 & 90* (Paris: L. Fournier).

Foley, Robert T. (2005), *German Strategy and the Path to Verdun: Erich von Falkenhayn and the Development of Attrition, 1870–1916* (Cambridge: Cambridge University Press).

Gascouin, Émile (1920), *L'Evolution de l'artillerie pendant la guerre* (Paris: Imprimerie Nationale).

Goya, Michel (2004), *La chair et l'acier: l'invention de la guerre moderne, 1914–1918* (Paris: Tallandier).

Goya, Michel (2008), *Le processus d'évolution tactique de l'armée française de 1871 à 1918* (doctoral thesis for University of Paris IV)

Griffith, Paddy (1994), *Battle Tactics on the Western Front: The British Army's Art of Attack, 1916–1918* (London: Yale University Press).

Gudmundsson, Bruce I. (1989), *Stormtroop Tactics: Innovation in the German Army, 1914–1918* (Westport, CT: Praeger).

Langlois, Hippolyte (1908), *L'Artillerie de campagne en liaison avec les autres armes* (Paris: Librairie Militaire R. Chapelot).

Langlois, Hippolyte (1909), *Lessons from Two Recent Wars: The Russo-Turkish War and South African Wars* (London: Mackie and Co.).

Laparra, Jean-Claude (2006), *La machine à vaincre, De l'espoir à la désillusion: histoire de l'armée allemande 1914–1918* (Mercuès: Imprimerie France Quercy).

Lloyd, Nick (2006), *Loos 1915* (Stroud: Tempus).

Lucas, Pascal Marie Henri (1925), *L'Evolution des idées tactiques en France et en Allemagne pendant la guerre de 1914–1918* (Paris: Berger-Levrault, third edition).

Millett, Allan R. and Murray, Williamson (1988), *Military Effectiveness, Volume I: The First World War* (London: Unwin Hyman).

Mosier, John (2001), *The Myth of the Great War: A New Military History of World War I* (London: Profile Books).

Palat, Barthèlemy Edmond (1927), *La grande guerre sur le front occidental* (Paris: Chapelot).

Percin, Général (1921), *Le massacre de notre infanterie 1914–1918* (Paris: n.p.).

Philpott, William (2009), *Bloody Victory: The Sacrifice on the Somme and the Making of the Twentieth Century* (London: Little, Brown).

Porch, Douglas (1981), *The March to the Marne: The French Army 1871–1914* (Cambridge: Cambridge University Press).

Porte, Rémy (2005), *La mobilisation industrielle: «premier front» de la grande guerre?* (Cahors: 14–18 Éditions).

Prior, Robin and Wilson, Trevor (1992), *Command on the Western Front: The Military Career of Sir Henry Rawlinson 1914–18* (Oxford: Basil Blackwell).

Rimailho, Émile (1924), *Artillerie de campagne* (Paris: Gauthier-Villars).

Sheffield, Gary (2001), *Forgotten Victory: The First World War: Myths and Realities* (London: Headline).

Sheffield, Gary and Todman, Dan (2004), *Command and Control on the Western Front: The British Army's Experience 1914–1918* (Staplehurst: Spellmount).

Sheldon, Jack (2005), *The German Army on the Somme 1914–1916* (Barnsley: Pen & Sword Books).

Stevenson, David (1996), *Armaments and the Coming of War: Europe, 1904–1914* (Oxford: Clarendon).

Memoirs, Doctrine and Unit Histories

Fayolle, Marie Émile (1964), *Cahiers secrets de la Grande Guerre* (Paris: Librairie Plon).

Foch, Maréchal (2008), *Oeuvres complètes, Tome I: Des Principes de la Guerre* (Paris: Economica).

Guiral, Louis (1965), *'Je les grignote ...': Champagne 1914–1915* (Paris: Hachette).

Humbert, Capitaine (1919) *La Division Barbot* (Paris: Librairie Hachette et Cie).

Joffre, Maréchal (1932), *Mémoires du Maréchal Joffre (1910–1917)* (Paris: Librarie Plon).

Official Histories

Chatinières, Paul, Janet, Paul, Lefranc, Abel and Landousie (1923), *Les armées françaises dans la Grande Guerre* (Paris: Imprimerie Nationale) [relevant for vols II and III].

Edmonds, Brigadier-General Sir James E. (1928), *Military Operations in France and Belgium, 1915* (London: Macmillan).

Biographies

Greenhalgh, Elizabeth (2011), *Foch in Command: The Forging of a First World War General* (Cambridge: Cambridge University Press).

Liddell Hart, Basil (1937), *Foch: The Man of Orleans* (London: Penguin).

Marshall-Cornwall, General Sir James (1972), *Foch as Military Commander* (London: Willmer Brothers).

Archive Catalogues

Nicot, Jean (1968), *Répertoire numérique: Des journaux des marches et operations 1914–1918* (Paris: Imprimerie Nationale).

Articles

Arnold, Joseph C. (1978), 'French Tactical Doctrine 1870–1914', *Military Affairs*, vol. 42, no. 2, pp. 61–67.

de Chasteigner, Lieutenant-Colonel (1965), '1915: Le martyre de l'infanterie. Un Exemple: Les Eparges', *Revue Historique des Armées*, vol. 21, no. 2, pp. 7–20.

Doughty, Robert A. (2003), 'French Strategy in 1914: Joffre's Own', *Journal of Military History*, vol. 67, no. 2, pp. 427–454.

Genevoix, Maurice (1965), '1915 «Année Terrible»', *Revue Historique des Armées*, vol. 21, no. 2, pp. 5–6.

House, Jonathan M. (1976), 'The Decisive Attack: A New Look at French Infantry Tactics on the Eve of World War One', *Military Affairs*, vol. 40, no. 4, pp. 164–169.

House, Jonathan M. (1984), 'Toward Combined Arms Warfare: A Survey of 20th-Century Tactics, Doctrine and Organization', *Combat Studies Institute*, Research Survey No. 2.

Ripperger, Robert M. (1995), 'The Development of French Artillery for the Offensive, 1890–1914', *Journal of Military History*, vol. 59, no. 4, pp. 599–618.

Index

Made in the USA
Monee, IL
13 September 2026

62976740R00111